PRACTICAL HANDBOOK

OF

DIGITAL MAPPING:

TERMS AND CONCEPTS

EDITOR-IN-CHIEF:
SANDRA LACH ARLINGHAUS, Ph.D.
Institute of Mathematical Geography

SPECIALIST ASSOCIATE EDITOR:
ROBERT F. AUSTIN, Ph.D.
Austin Communications Education Services, Inc.

ASSOCIATE EDITORS:
William C. Arlinghaus, Ph.D., Lawrence Technological University
William D. Drake, Ph.D., The University of Michigan
John D. Nystuen, Ph.D., The University of Michigan

with assistance from:

Christine Kolars, M.P.H.
Kris Oswalt, M.B.A.

CRC Press
Boca Raton   Ann Arbor   London   Tokyo

COVER MAP AND PLATES I THROUGH XI

The cover map, based on a map provided by Environmental Systems Research Institute, Inc., shows a practical, graphic representation of assigning students to existing schools; conceptually, it is a structural model depicting this assignment as graph-theoretic stars.

PLATES XII THROUGH XVI

MAPS IN THE TERMS AND CONCEPTS PART

These maps were made by S. Arlinghaus using MapInfo (licensed to W. Drake/CSF) on a 486 Personal Computer with 16 Megabytes of RAM and a 240 Megabyte hard drive. The text was prepared by her on the same computer using MicroSoft Word for Windows, version 1.1 (licensed to W. Drake/CSF).

Library of Congress Cataloging-in-Publication Data

Practical handbook of digital mapping : terms and concepts / editor-in
 -chief Sandra Lach Arlinghaus ; specialist associate editor, Robert
 F. Austin ; associate editors, William C. Arlinghaus, William D.
 Drake, John D. Nystuen, with assistance from, Christine Kolars, Kris
 Oswalt.
  p.   cm.
 Includes bibliographical references and index.
 ISBN 0-8493-0131-9
 1. Digital mapping.   2. Geographic information systems.
 I. Arlinghaus, Sandra L. (Sandra Lach)
 GA139.P73   1994   93-23662
 526'.0285--dc20   CIP

PREFACE

Anyone who already uses maps will benefit from using digital maps; many more who do not presently use maps, but who do use tables or arrays of data in problem solving, would also benefit from using digital maps.

This handbook is a reference work explaining a selection of the highly technical language, both specific terms and abstract concepts, of digital mapping. Digital maps--those produced and stored in a computer--have many advantages over manually-drafted maps, especially in situations of frequent growth or change. Digital maps are produced from computerized data bases which may be queried and used by several people simultaneously. Some of the advantages of digital maps are: ease and speed of revision, inexpensive production of short-run special purpose maps, high level of precision in representation, and outstanding capabilities for spatial analysis. Changes in the underlying database produce corresponding changes in the map.

In 1988, the country of Qatar (population over 400,000, situated on the Arabian Gulf) initiated a nationwide digital mapping program to meet the needs of the country for computer-assisted record keeping, map keeping, and planning. Government workers in many other countries are finding it necessary to learn about GIS, as are business people, students, educators, and researchers. This handbook is designed as a practical guide to digital mapping--for beginner and experienced digital mapper alike.

The sets of terms and concepts that are related to digital mapping are drawn from fields as far-flung as: geography, urban planning, intelligent vehicle highway systems, transportation planning, natural resources, public health, population planning, operations research, engineering, mathematics, computer science, telephony science, fiber optic cable technology, GIS and AM/FM business arena, railroads, aerial surveying, remote sensing, photography, and projects in the international development community. This book will therefore assist individuals with interests related to any combination of those fields. The broad range of topics treated reflects the wave of the future, as for example in integrating satellite receivers into laptop computers or in using widely spread networks to disseminate computing capability.

Useful Handbook Features:
1. The comprehensive character of the book draws on a mix of business, academic, and the international development's community approaches to digital mapping.

2. Alphabetical organization of terms and concepts provides quick and easy reference, even in the field.

3. Strong emphasis on basic map making principles makes concepts and practicality merge easily. All graphic displays are digital maps. The color plates (I to XI) have their own index cross-referenced to the Terms and Concepts of Part I, so that readers attracted by the graphic display might learn easily where to read about it.

4. A case study from the international development community illustrates a use of digital mapping by UNICEF (Plates XII to XVI). The case study has an index of its own, which should help in coordinating this material with that of Part I.

5. A unique collection of references to the GIS literature is included. These references have been drawn together over some years, and many of the articles are not easily found in computerized databases. Individually numbering these references made possible a separate index to the references alone (cross-referenced to Part I). This index should provide quick entry into the literature.

As is necessarily the case with this sort of endeavor, it needs immediate and continuous updating--as this first edition goes to press, a second is in the works. If you find something that you wish to share with us for the next edition--new material, corrections, or whatever--please communicate with:

William C. Arlinghaus
ATTN: CRC Handbooks
Department of Mathematics and Computer Science
Lawrence Technological University
21000 West Ten Mile Road
Southfield, MI 48075
Arlinghaus@LTUVAX

Thank you in advance; we have tried to be careful and to contribute something that is useful and different; in the end, despite all the care of the many who have generously offered time, effort, and advice, the blame for errors, omissions, or poor judgment must rest with the Editor-In-Chief, alone.

Sandra Lach Arlinghaus, Ann Arbor, MI   September 3, 1993.

ABOUT THE EDITORS

Except where noted in the text or the Acknowledgments, this book was written primarily by the Specialist Associate Editor, Robert F. Austin and by the Editor-In-Chief, Sandra Lach Arlinghaus. The role of the board of Associate Editors was to review and critique the material of these two author-editors and guide the direction of the document: in terms of general scope, graphic presentation, and accuracy of detail. In addition, W. Arlinghaus wrote a number of the entries in the "terms and concepts" part and devised the key-word indices; W. Drake found the case study and helped to coordinate book production; and, J. Nystuen wrote the majority of the case study and helped to coordinate the graphics.

The set of five editors has combined experience of well over 100 years in various disciplines and activities related to digital mapping, including an extensive list of publications and invited/contributed lectures, at the professional level. Areas in which they have professional expertise include: geography (mathematical, spatial analysis, cartography, GIS, cultural), regional science, urban planning, intelligent vehicle highway systems, transportation planning, natural resources, public health, population planning, operations research, engineering, mathematics (graph theory, combinatorics, modern algebra, non-Euclidean and Euclidean geometry), computer science (theoretical), computer science in the automotive industry, telephony science, telephone lineman, fiber optic cable technology, GIS and AM/FM business, aerial surveying and photography, and field study in the international arena.

Sandra Lach Arlinghaus, Ph.D. Geography, University of Michigan.
Founding Director, Institute of Mathematical Geography.

Robert F. Austin, Ph.D. Geography, University of Michigan.
President, Austin Communications Education Services, Inc.

William C. Arlinghaus, Ph.D. Mathematics, Wayne State University.
Professor and Chairman,
Department of Mathematics and Computer Science,
Lawrence Technological University

William D. Drake, Ph.D. Operations Research, University of Michigan.
Professor,
School of Natural Resources, School of Public Health, and,
College of Architecture and Urban Planning,
University of Michigan.

John D. Nystuen, Ph.D. Geography, University of Washington.
Professor,
Geography and Urban Planning,
College of Architecture and Urban Planning,
University of Michigan.

The research for this book was drawn from the field experience of the editors and from background reading listed in the references section; editor Robert F. Austin has been accumulating terms and concepts from the GIS, AM/FM, telephony business, aerial survey business, railroad industry, and others for the past 15 years. These written documents, developed in partial collaboration with Robert Austin, served as the base for this book. Editor S. Arlinghaus added to it from her direct teaching experiences in geography, mathematics, and natural resources, and from her various research interests. A number of the terms and concepts are unique to this book insofar as they suggest directions for theoretical research projects linking mathematics and digital mapping. Many of the others are more practical.

We owe a particular debt to numerous colleagues who have gone before us in bringing order to the world of GIS terminology, digital mapping, medical imaging, low dimensional topology, and others. We offer this document as a modest block resting on their solid foundation, in the hope that it will supplement their fine materials and be of use to students, researchers, business people, and others in a variety of practical contexts.

ACKNOWLEDGMENTS

One of the nice parts of writing a book of this sort is having the opportunity to get advice from a diverse set of individuals. In addition to the editorial board, we thank the following individuals for their written contributions to this work.

Robert Austin, CEO of Austin Communications Education Services, Inc., contributed original material to many of the terms involving fiber optic cable and related telecommunications concerns. He also helped in reviewing the final document. Long a pioneer in the fiber optic cable world, one of Robert's practical achievements involved the design of the first North American fiber optic cable system actually used to transmit a live sportscast (the Tampa Bay Buccaneers on Monday Night Football in 1973).

Frank Harary, Distinguished Professor of Computer Science (New Mexico State University), Professor Emeritus of Mathematics (University of Michigan), and formerly Instructor of Physics (Princeton University) contributed written definitions of the terms "graph theory," "hypercube," and "structural model." Always an enthusiast for even more applications of graph theory, Mr. Graph Theory (as Frank is affectionately known to many) is a constant source of encouragement and inspiration. Indeed, those who have not yet met Frank might get to know him through any of his more than 600 published journal articles or 6 books.

Others have offered substantial support and guidance in less tangible, but no less important, ways.

Kenneth H. Rosen of A.T.&T. Bell Laboratories first encouraged Sandra Arlinghaus to consider the idea of writing handbooks and put her in touch with Wayne Yuhasz at CRC Press. How suitable it is that the practical experience that Ken and Sandy shared as Lecturers of Mathematics at The Ohio State University should eventually turn into a practical handbook!

Frederick L. Goodman of the School of Education of The University of Michigan has been an enthusiastic supporter of this venture and of the broader idea of education and GIS--from state of the art technology to state of the art education and back again. Fred's cleverness with educational games is turning games into maps and maps into games.

A large part of digital mapping involves the business world of GIS and AM/FM, and the institutional world beyond university walls.

Environmental Systems Research Institute, Inc. (ESRI) has been an invaluable source of digital images, and has served as an outstanding model for the sort of cooperation that can occur between the academic and commercial worlds. We thank, especially, Carl Sylvester of ESRI. We also thank the many others at ESRI who were very helpful in assembling a wide range of attractive and pertinent digital images.

Community Systems Foundation (CSF) of Ann Arbor has generously provided access to both human and software resources for this project; thanks to W. Drake and K. Oswalt (both of CSF) for putting us in touch with J. Sherry of UNICEF. Christine Kolars and Kris Oswalt have shared their expertise in the field of international health; work they have written bears their names in this book. We thank them, and CSF, for their outstanding effort.

During the course of the past two years, during which this physical document was written, there are many with whom we have had conversations. To those friends, and colleagues who have written a brief e-mail note of encouragement, noted a related newspaper article, suggested a set of acronyms to include, helped with proofreading, offered advice on technical printing matters, or other similar comments, we offer our thanks. In particular, we thank William E. Arlinghaus, Michelle Austin, Nanette R. LaCross, and Gwen L. Nystuen, for their patience and generosity.

Finally, Wayne Yuhasz, Executive Editor--Physical Sciences at CRC Press, has made all of this possible. Wayne is delightful to work with; his good ideas, patience, and broad understanding of the sciences have made the writing of this book a pleasure.

TABLE OF CONTENTS

PART I: DIGITAL MAPPING, TERMS AND CONCEPTS, ALPHABETICALLY ARRANGED

A a

A Programming Language, APL

high-level programming language used on IBM mainframes and PCs for scientific and mathematical applications. A structured programming language.

AAG

see Association of American Geographers.

abandoned cable

telephone cable that is in place but abandoned.

abbreviated addressing

packet-switched network addressing scheme in which simple mnemonic code is used instead of complete addressing information. The cross reference to the complete address is stored in the Packet Assembler/Disassembler (PAD).

abbreviated dialing

speed telephone dialing.

abort

an exception that occurs in the middle of a computer instruction and potentially leaves the computer's registers and memory in an indeterminate state, such that the instruction cannot necessarily be restarted.

ABR

see Automatic Baud Rate detection

abscissa

mathematically, horizontal (x-axis) coordinate in a Cartesian graph.

absolute accuracy

see accuracy.

absolute index mode

an indexed addressing mode--the base operand specifier is addressed in absolute mode.

absolute mode

in absolute mode addressing, the PC is used as the register containing the address of the location of the actual operand.

absolute time

absolute time values expressing a specific date (month, day, year, and time of day) are always given in the computer system as positive numbers.

absorption

process in which radiant energy is absorbed and changed into other forms of energy; processes involving energy are important in satellite systems gathering data about the earth.

AC

see Alternating Current.

ACA

see American Cartographic Association.

Academy of Sciences (France)

the French Academy was founded in the last half of the 17th century; among its chartered purposes was the improvement of charts for navigation. The solution to navigating effectively in the open seas lay in finding an accurate, practical, method for measuring longitude at sea. Cartography was an enduring concern, not only for navigation purposes but also for determining the shape and size of the earth and the shape and position of coastlines and other boundaries. These are enduring issues that are intimately associated with our ability to measure time; as clocks have improved, so have earth measurements, down to the present, with atomic clocks governing the accuracy of timing alignments in systems of satellites.

acceptance angle

the maximum angle, measured from the longitudinal axis of an optical fiber to an incident ray, at which total internal reflection of the incident ray occurs. A typical acceptance angle is 20 degrees.

acceptance cone

a right circular cone whose apex angle is equal to twice the acceptance angle. Rays of light within this cone can be coupled into the end of an optical fiber--total internal reflection is still maintained for all rays in the cone. A typical apex angle is 40 degrees.

acceptance test

evaluation of recently acquired system's performance or data quality.

access

used as a verb by computer scientists--to retrieve a file or program from an on-line computer device or storage.

access group

in a Local Area Network, all stations that have identical rights to make use of resources (such as computer, network, or Private Automatic Branch Exchanges--PABX).

access method

a host program, in an IBM computing environment, that manages the communication of data between main storage and an input/output device; Basic Telecommunications Access Method (BTAM), Telecommunications Access Method (TCAM), and Virtual

Telecommunications Access Method (VTAM) are common access methods.

(in Local Area Network technology) a procedure to allow LAN stations to make use of ("access") the network's transmission system. This access is either shared access or discrete access. Shared access is further subdivided as explicit or as contended access.

access mode

one of four computer processor access modes in which software executes; processor access modes are, in order from most to least privileged and protected: kernel (mode 0), executive (mode 1), supervisor (mode 2), and user (mode 3); when the processor is in kernel mode, the executing software has complete control of,   and responsibility for, the system; when the processor is in any other mode, the processor is inhibited from executing privileged instructions; the operating system uses access modes to define protection levels for software executing in the context of a process (DEC).

access path

description of the chain of directories defining the location of a file within a logical hierarchy in a computer system.

access rights

level of interaction permitted with stored data files; e.g., read-only.

access technology

technology   to   provide   users   with   physical   access   to telecommunications services.

access time

the time delay required for reading data from or to an external device such as disk drive.

access type

the ways in which a processor accesses instruction operands or in which a procedure accesses its arguments. Access types include: read, write, modify, address, and branch.

access violation

an attempt to reference an address that is not mapped into virtual memory or an attempt to reference an address that is not accessible by the current access mode.

accessibility

extent to which it is easy to gain access. In Plate I, the central areas have a high degree of access to the demanded retail services.

account name

a string of alphanumeric characters that identifies a particular computer account; typically, the name is the user's accounting charge number, rather than the user's internal identification number.

accunet

a set of AT&T high-speed digital facilities, including packet-switched network, Dataphone Digital Service (DDS), and formatted T1 (of 1.544 or 2.048 Mbps bandwidth) offerings.

accuracy

the degree of conformity with a recognized standard, or the degree of correctness of a measurement. Accuracy concerns the quality of an answer; precision concerns the quality of the procedure which led to the answer. In GIS environments, "accuracy" often refers to photogrammetry and Global Positioning Systems concerns. An accuracy of plus-or-minus 10 meters means that any location on the map is correct within 10 meters of actual ground measurement.

absolute accuracy

accuracy of position of individual map features compared to real-world (terrestrial) location of the feature.

compilation accuracy

in the compilation of a map as a sequence of overlays, the final product is only as accurate as the margin for error in each of the overlays, whether or not symbol and line placement is manual or digital. The use of a computer does not guarantee accuracy in compilation; layers prepared using a computer may fail to register, even though registrations may be perfect along tick marks. The reasons for such failure are myriad; they may be due to inaccuracies in the underlying data set, or to variation, from layer to layer, in definition of the same terms. The map compiler needs to execute the same level of care in compiling computer maps as in manually-drawn maps.

database accuracy

accuracy of content, accuracy of spatial and topological relationships when expressed in numbers.

mapping accuracy

proportion of measured features that must meet some a priori criteria; generally expressed as acceptable error.

relative accuracy

accuracy of position of individual map features compared to   other positions of other features on the same map.

ACK

see acknowledgment.

Acknowledgment, ACK

control character used with Negative Acknowledgment (NAK) in Binary Synchronous Communications (BSC) protocol. It shows that the previous transmission block was correctly received and that the receiver is now prepared to accept the next block. ACK is also used in a similar fashion in the ETX/ACK method of flow control and in Hewlett-Packard's ENQ/ACK protocol.

acoustic coupler

device that converts electrical into audio signals, and vice-versa. It enables data transmission over public telephone networks via a telephone handset, as a modem.

acronym

word formed from first (or other) letters of a grouping of words. For example, ASCII is an acronym for American Standard Code for Information Interchange.

ACS

see Advanced Communications Service.

ACSM

see American Congress on Surveying and Mapping.

active system

in remote sensing, a system that transmits (its own) electromagnetic signals and then records the energy that is reflected or refracted back to the sensor.

Actual Measured Loss, AML

the actual reading from a transmission measuring device (cf. EML).

acuity

visual acuity refers to the ability of the human eye to resolve detail; generally, distinctions in detail are easier to make on monochromatic backgrounds than on colored backgrounds. When colored backgrounds are used, those that mix a variety of wavelengths (such as brown) are bad choices.

ADA

high-level programming language used for military programming applications (developed by DOD). A structured language that encourages the use of modular programs (as does Pascal, for example). Optimized modules are spliced together to obtain an entire program.

adaptive equalizer

an automatic (generally) equalizer that adjusts to meet varying line conditions.

adaptive sampling

intelligent sampling procedure in which the experience of previous sampling efforts guides continuing sampling efforts.

ADCCP

see Advanced Data Communications Control Procedures.

additional facilities

in packet-switched networks--as contrasted with essential facilities--network facilities chosen for a particular network.

additive color
see color.

address
(noun)   the unique designation in a computer memory for the location of data or for the identity of an intelligent device. Additional devices on one communication line must have unique addresses so that each might receive messages correctly--as a one-to-many transmission.

(verb) to address a message means to include on it a code showing the location of the desired receiving device.

address access type
a command in which the specified operand of an instruction is not directly accessed by the instruction.

address bus
the communication lines used to transfer address information between components in a computer system.

address geocoding
alignment of municipal or other addresses with spatially related information. For example, an address might be related to sewer service lines which have been coordinatized so that the address is matched to the appropriate segment of sewer (from an underlying database) that is physically under the street.

address matching
relating street addresses to geographic units such as census tracts.

address space
the set of all possible addresses available to a process; virtual address space refers to the set of all possible virtual addresses; physical address space refers to the set of all possible physical addresses.

addressability
the number of pixels in the product of the x- and y-axes on a display screen (x-axis across the top, and y-axis down the left side).

addressable point
position on a display screen that can be specified uniquely as a coordinate pair.

addressable space
memory size of a display board which may be greater than or equal to the area of a display screen.

adjacent
two nodes in a graph are said to be adjacent if there is an edge joining them.

adjoining sheets
two or more maps, whose content is continuous across their adjacent boundaries at corners or edges of the map sheets. Plate II illustrates this idea; note that on it, three different images are shown--left side, and

right side (top and bottom)--and that the content is continuous across the boundaries.

Advanced Communications Service, ACS
    AT&T's proposed packet-switched network.

Advanced Data Communications Control Procedures, ADCCP
    the USA's Federal Standard communications protocol.

Advanced Mobile Phone System, AMPS
    analog, cellular mobile phone system developed by AT&T.

Advanced Program-to-Program Communications, APPC
    IBM protocol that allows peer-to-peer communications.

Advanced Research Projects Agency NETwork, ARPANET
    (Department of Defense) a wide-area packet network that was funded by DARPA; the ARPANET is composed of packet switch nodes connected by leased lines. ARPANET is the first major network of this sort; it connects about 150 sites doing research in for the U.S. federal government at 150 universities and corporations. It is contained within Internet.

Advanced Very High Resolution Radiometer imagery, AVHRR
    imagery produced by NOAA satellites.

Advanced Visible/InfraRed Imaging Spectrometer imagery, AVIRIS
    images collected by NASA aircraft. Plate III shows a multispectral satellite image of San Francisco Bay.

aerial cable or wire
    the type of construction in which wire or multipair cables are suspended from utility poles. Aerial construction is usually used for distribution networks, but may also be used for feeder or toll networks. The portion of the aerial cable which is placed in a building is called house, building, or riser cable.

aerial photography
    method of photographing from an aerial platform. The result is a photograph of land taken from the air, generally vertical but occasionally oblique. Aerial photographs might be printed on paper or on transparent film; a common size is nine inches by nine inches. Vertical photography (orthophotography) is used for photogrammetry--a high degree of accuracy is required. Oblique photography is generally used for more general purposes, and thus does not require as high a degree of accuracy as orthophotography. One special purpose of oblique photography is to verify certain attributes. The photo in Plate III is taken from an oblique angle.
    Generally, aerial photography has been, and is, important in a variety of military contexts. It is also critical to a variety of agencies such as the U.S. Geological Survey for topographic mapping and the U.S. Forest Service for inventories of timber.

Aerial Photography Summary Record System, APSRS
aerial photography cataloging system administered by NCIC, USGS.
aerial survey
survey information take from aerial photographs or from remotely sensed data using infrared, gamma, or ultraviolet bands of the electromagnetic spectrum.
affine
affine geometry was first recognized by Leonhard Euler in the eighteenth century. It is a geometry that can be extracted from Euclidean geometry. Euclid's fifth postulate is dominant in affine geometry. This postulate says that through a given point not on a given line, there is a unique line that can be drawn through the given point that is parallel to the given line. The affine theorems in Euclid are those which remain invariant under parallel projection from one plane to another. There have been a variety of uses for various geometries, other than the purely Euclidean, over the past few centuries in various disciplines. The affine propositions are true not only in Euclidean space but also in Minkowski's geometry of time and space. The interested reader should examine works, among others, of William Kingdon Clifford (19th century) and H. S. M. Coxeter (20th century) for theoretical developments.

In a digital mapping context, affine projections are important when one views layers of a map as parallel projections from one plane to another. They are also important in singling out a map feature to look at more closely, or to "zoom-in" on. Thus, the geometer's transformations of translations, half-turns, and dilatations, become the mapper's translations, rotations, and scaling transformations. The mathematician finds it important to study under what condition these transformations form a closed mathematical structure; the mapper realizes this closure in being able to get back to the starting point after a sequence of mapping transformations. Examples of affine transformations are given below.

Translation: a translation is a parallel displacement--there are no points left invariant under translation. Literally, translation refers to "sliding across"--thus, linguistically "porte" is displaced to "door"--as a parallel expression for a familiar physical object. Thus, a square in the plane, slid to another location, or address, in the plane, may have a location different from the original, but the physical object of "square" remains the same. The idea of translation is independent of the means for assigning an address to the position of the geometric object, just as the concept of "door" is independent of how it is spelled in various natural languages. Often coordinate systems are introduced as the mathematical languages in which to spell-out or measure translation;

they can be useful when well-chosen, but they are not required. When coordinate systems are employed, translation of a geometric object can be expressed in terms of translation, or movement to a new location, of the coordinate system.

Half-turns and rotations: a half-turn is a rotation of a line segment through pi. An axiomatic development of affine geometry sees translations, half-turns (and not arbitrary rotations), and dilatations as fundamental transformations from which others are formed. Thus, the general rotation emerges as a product of translations. Many digital mapping books list a rotation as an affine transformation without regard to how it was conceived. At present, this distinction causes no difficulty; but, it could.

When a coordinate system is present, the axes are rotated to measure the extent of change in position. A rotation has one invariant point-- often thought of as the intersection point of the axes of the rotation.

Dilatation or scale transformation: shrinking or enlargement of all or some of a geometric figure. When coordinates are used, the rescaling of units along the axes measures the scale transformation.

These affine transformations can all be compactly represented, in the presence of a coordinate system, using matrices and theorems from linear algebra. The order in which transformations are composed, or multiplied as matrices, is significant. In some cases the multiplication is commutative; in other cases it is not. The product of two translations is once again a translation and the order in which the product is formed does not matter. The set of all translations forms a mathematical group. There are many interesting results from group theory that might well be useful in developing digital mapping based on affine (or other) transformations.

The surface of the theory of affine transformations has, to date, only barely been touched by digital mapping products. There are various theorems from affine geometry that might lead the individual interested in mapping geometry in a number of different directions. As with any mathematical structure, however, the researcher must work through the ordered sequence of theorems and related material to understand how to apply results in a reliable manner. One early theorem in affine geometry is Pick's Theorem:

Given a simple polygon whose vertices are lattice points in the plane. Its area, A, is a function of b, the number of lattice points on the polygon boundary, and c, the number of lattice points strictly within the polygon--

$$A = 0.5b + c - 1.$$

Thus, in any rectangular lattice, for example, a unit triangle has (by Pick's Theorem) area of 0.5 of a unit lattice rectangle (0.5 = 0.5*3 + 0 - 1), as one would require naturally.

Beyond affine geometry, there is little use made, in digital mapping, of geometries that do not obey Euclid's postulate of unique parallels; this is a fertile source of research projects that is virtually untapped, as is associated material concerning group theory.

agglomeration

one method of cartographic classification in which clusters of points are represented by a single typical point. The same sort of strategy may be employed with lines or areas of a map. Determining what is meant by "typical" is part of the cartographer's art which may be aided by statistical and computational tools. Plate IV shows agglomerations of a school age population.

aggregate input rate

sum of all data rates of terminals and ports connected to a multiplexor or concentrator. The modifier "burst" refers to the instantaneous maximum of this rate.

AGS

see American Geographical Society.

AI

see Artificial Intelligence.

air dryer

a piece of equipment that supplies the source of dry air to telephone cables. It is usually located in a central office, but is sometimes remotely located.

air mass

a parcel of air of generally uniform temperature and humidity that can move as a meteorological unit.

air pap

a computer program that analyzes and designs outside plant pressurization systems.

air slide

aerial 35 mm slide with orthogonal, downward, view.

air video image

same as airslide, but with a video camera and recorder.

air-pipe

a piece of tubing used to route the supply of air pressure from the CO dryer to the manifolds. Symbolized PP (pressure pipe).

airbrush

pencil-shaped ink atomizer used to draw fine detail and to create smooth tones. It is powered by compressed air.

albedo

a measure of the portion of incident electromagnetic radiation reflected by the earth's surface (or other surfaces); expressed as a ratio of the light reflected to the light received. Albedo affects photographic emulsions, eyes, and various other receptors--a consideration when using photographs to map landforms. White pin-points in the highly urbanized area on Plate III suggest points with a high level of reflectance.

Albers' conic projection

an equal-area (conic) projection, based on two standard parallels along which there is no angular distortion, that is best suited to mapping an area whose east-west extent is substantially greater than its north-south extent, (particularly in middle to high latitudes). It is thus well-suited to mapping the conterminous United States and is often employed in that capacity. See map in Figure 1.

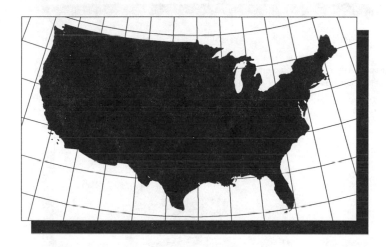

Figure 1. Albers conic equal area. Standard parallels positioned to make the co-terminous U.S.A. appear true.

When the entire globe is viewed in this manner, on a cone unrolled in the plane, with the apex of the cone situated so that the standard parallels fall in a position to make the U.S.A. look a reasonable shape, so too do other landmasses at about the same latitude. Figure 2 shows clearly how this map is developed from a cone. The farther one moves away from the region of standard parallels, the greater the distortion. Note in Figure 2 that Alaska has east-west extent that is too great, and that landmasses in the southern hemisphere almost wrap under the spherical suggestion of the graticule.

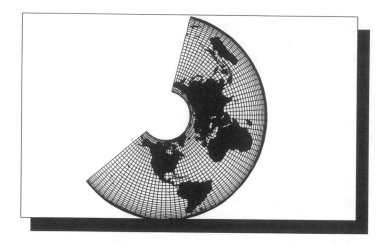

Figure 2. Albers conic equal area. Standard parallels positioned to make the co-terminous U.S.A. appear true--much of cone shown.

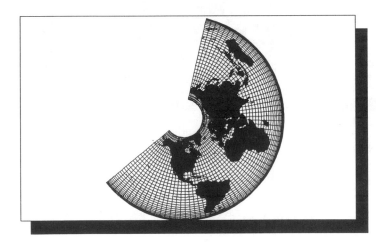

Figure 3. Albers conic equal area. Standard parallels positioned to make North America appear true--much of cone shown.

If one wished to broaden the belt of the standard parallels, in order to get more of North America closer to the best position, then one might pull the cone apex farther into space; Figure 3 shows the result of doing so to the cone. When the co-terminous U.S.A. is zoomed-in on from this vantage point, the map still looks quite good, although there is some compression in the east-west extent (Figure 4; compare Figure 4 to Figure 1).

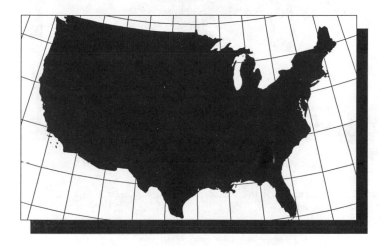

Figure 4. Albers conic equal area. Standard parallels positioned to make North America appear true. Compare to Figure 1.

If instead, one wished to improve the shape of Alaska, then the apex of the cone might be pulled even farther into space to reposition standard parallels to favor focusing on Alaska (Figure 5). This projection shows circumpolar regions with little distortion; it would be a poor choice, however, for displaying Australia. Indeed, even the co-terminous U.S.A. is not close enough to the standard parallels; it is badly misshapen (Figure 6).

If one wished, instead to move the apex of the cone closer, to position the standard parallels to focus on Hawaii, only a narrow band of low-latitude landmasses appear true (Figure 7); this projection is not well-suited to focusing on the middle of the sphere (hence the comment above about mid-latitude and high-latitude uses). When the coterminous U.S. is viewed from this perspective unacceptable stretching is evident (Figure 8).

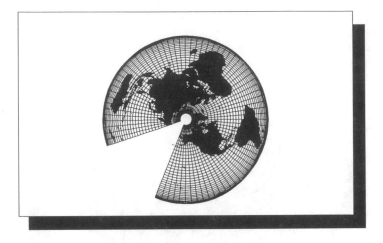

Figure 5. Albers conic equal area. Standard parallels positioned to make Alaska appear true--much of cone shown.

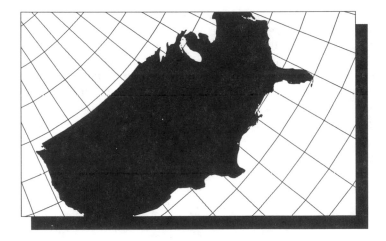

Figure 6. Albers conic equal area. Standard parallels positioned to make Alaska appear true.

The projection one chooses can make a difference. Indeed, the way in which the projection is positioned can also make a difference. See also, conic projection.

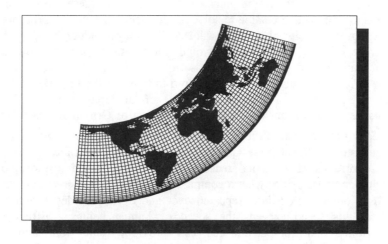

Figure 7. Albers conic equal area. Standard parallels positioned to make Hawaii appear true--much of cone shown.

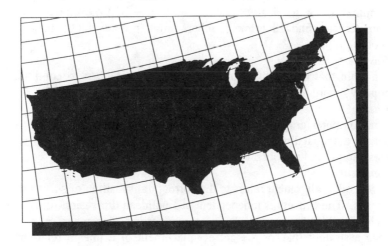

Figure 8. Albers conic equal area. Standard parallels positioned to make Hawaii appear true.

Algemeine Zeichen Program, AZP

a type of graphic command (a protocol).

algorithm

a definite procedure for solving a problem using a finite number of steps. The steps should be precisely defined in an ordered sequence. An algorithm should be able to solve all problems of a given form, not just a problem with given input values. Thus a procedure to find the average of 2, 3, 5, 7, and 11 would not be an algorithm. But a procedure to find the average of a set of any five integers would be. The efficiency of algorithms is measured in terms of the time they take to solve the problem relative to the size of the input data set. For instance, an algorithm is said to be of order n or O(n) if its speed is directly proportional to the number n of input values; it is $O(n^2)$ if its speed is proportional to the square of the number of input values.

Algorithms have gained renewed prominence in the age of the digital computer, since what a computer does best is fast computation, and an ordered procedure, repeated over and over, provides a setting where this speed exceeds the speed of human beings. Efficient algorithms and computer speed can provide results that human computation could not do in years. One simple example is the Sieve of Eratosthenes, an ancient procedure which describes how to determine which numbers up to a certain size are prime numbers (a prime number is a positive integer greater than 1 whose only divisors are 1 and itself). A human being can probably find all prime numbers smaller than 300 in ten minutes. A good algorithm for the Sieve of Eratosthenes and a computer can find all prime numbers smaller than 10,000 in a few seconds.

alias file

a table that relates the attributes of an item of plant to continuing property record (or some other record) item numbers.

aliasing

in computer graphics, in a raster environment, the jagged staircase look to diagonal lines formed from a sequence of square pixels with sides parallel to horizontal and vertical axes.

alignment

generally, alignment refers to the correct transposition of survey data to maps. The concept is independent of Euclidean dimension--thus, one might consider an interpretation as the correct placement of points along a straight line, or the correct placement of a line in relation to other lines. In Plate V, utility lines are aligned with parcel boundaries. Plate VI also shows alignment of areas--red lines cross aligned tiles.

allocation

apportionment for a specific purpose. Plate IV shows the assignment of locations by color. Plates VII and VIII both illustrate the allocation of the school population relative to bus routes (Plate VII) and distance to school (Plate VIII). The Cover Plate shows assignment of origins (home places) to nearest destinations (schools) with only a few exceptions, observable where lines of the spiders/stars cross.

allocation area

the fundamental geographic unit for administering feeder facilities and for monitoring problem data in the Distribution Plant.

ALOHA

University of Hawaii experimental packet-switched network (1970s) implemented on radio.

alpha test

preliminary version of software.

alphanumeric

term describing a character set containing letters ("alpha"), numerals ("numeric"), and various punctuation and other marks. Telephone books or mailing addresses are examples of traditional alphanumeric lists.

alphanumeric character

an upper or lower case letter (A-Z, a-z), a dollar sign ($), an underscore (_), or a decimal digit (0-9).

alternate routing

in PABX technology--redundant routing enabling the completion of calls when the primary circuit is not available.

Alternating Current, AC

(electrical terminology) as opposed to DC, direct current.

altimetry

the technology of measuring altitudes; and, interpretation of altitude measurements.

altitude

vertical distance of an object or point above datum, usually mean sea level.

altitude tinting

a method of depicting the earth's surface on a map using hue, pattern, or value to "color" the regions between isometric lines, such as contour lines. Also called hypsometric tinting (coloring) or layer coloring.

ALU

see Arithmetic Logic Unit.

AM

see Amplitude Modulation; automated mapping.

AM/FM
   see Automated Mapping/Facilities Management.
American Cartographic Association, ACA
   a member organization of the American Congress on Surveying and Mapping (ACSM).
American Congress on Surveying and Mapping, ACSM
   Professional organization which, in 1982, at the request of the USGS, formed the National Committee on Digital Cartographic Data Standards (the "Moellering Committee").
American Geographical Society (of New York), AGS
   the oldest geographical society in the U. S. A., founded in 1851. The AGS Library, housed at the University of Wisconsin, Milwaukee, is, in many respects, one of the most extensive collections of geographical materials in the Western Hemisphere.
American National Standards Institute, ANSI
   American National Standards Institute (formerly the American Standards Association).   Voluntary organization.   Defined ASCII. Members include common carriers and standards organizations, such as IEEE, as well as manufacturers.   ANSI produces Federal Information Processing Standards (FIPS) for the Department of Defense.
American Standard Code for Information Interchange, ASCII
   an eight-bit code consisting of one parity bit and seven character code bits forming the 256 characters of a computer's basic alphabet. It was established by the American National Standards Institute to provide compatibility   between   data   services;   commonly   used   for communication between a CPU and peripherals; also referred to as USASCII. Equivalent to ISO 7-bit code. ASCII values 0 to 127 are standardized and include all the characters commonly used on a PC; generally, PCs use an extended ASCII character set for ASCII values 128 to 255.
AML
   see Actual Measured Loss.
amplifier
   electronic device used to boost the performance (measured in deciBels) of signals.
amplitude
   the maximum value of a varying wave form.
amplitude distortion
   undesired change in signal amplitude--often caused by non-linear elements in the communications path.
Amplitude Modulation, AM
   one method for adding information, by modifying the amplitude, to a sine wave signal.

AMPS

see Advanced Mobile Phone System.

analglyph

one way of simulating a stereo view; identical images in different colors are placed next to each other (as translations) with a two degree separation in viewpoint. If the left image is plotted in green and the right image in red, then the stereo view, producing a three dimensional visual effect, is achieved by viewing the paired images through glasses with a red left lens and a green right lens.

analog (analogue)

continuously variable quantities in which a numerical quantity is represented by a physical variable such as amplitude levels, phase, frequency, shape, or position, as opposed to discretely variable quantities or "digital" entities.

analog data

lower resolution data, often as an operational back-up for some higher resolution data. Form of transmission employing a continuous electrical signal (as opposed to a pulsed or digital signal) that varies in frequency or amplitude.

analog image

visible two-dimensional map.

analog loopback

a diagnostic test forming a loop at the interface of the modem with the telephone line.

analog to digital conversion

conversion of data from analog to machine usable digital format. Processes such as scanning or digitizing enable this conversion.

analog transmission

transmission of information using analogue, as opposed to digital, coding. Public telephone networks have typically employed analogue transmission systems with mechanical switching--prior to computerization. Digital transmission holds great promise for the transmission of video images and other technological advances.

analytical cartography

approach to cartography and to decisions involved in the making of maps that focus on the theoretical and mathematical underpinnings of the mapping process (see, Tobler, W. R., 1959; 1976).

analytical hill shading

a method of shading, using the computer, to show variation in terrain. A point source of light (or more than one) is assumed off the map; the illumination value is noted in every grid cell (or in every triangle in a triangulated irregular network) of the map to achieve the shading. See Plate I.

analytical triangulation

a technique in photogrammetric mapping that determines mathematical relationships among a set of control points. It uses computer aided interpolation of the control points in order to create a control network sufficiently dense to provide a positionally accurate framework for the photographic efforts to come. See Plate I.

anchor

a stationary device in the ground to which a guy is attached to support a pole.

anchor guy

guy and anchor combined. Extends from ground to pole.

anchoring stimuli

legend values on a map chosen to guide the map user into making good estimates of the values of mapped data from the legend.

ancillary data

supplement to a data set.

Angstrom, Å

unit of measurement, 10^{-10} meters--one ten-millionth of a millimeter. Used to measure wavelengths of light or X-ray; important in cartography when hues are employed in spectral progression.

angular field of view

see Field Of View.

annotation

on a map, the text or labels describing map features--street names, place names, and many others.

ANSI

see American National Standards Institute.

ANSI Technical Committee on Computer Graphics

"Moellering Committee" to establish graphics standards in analytical and computer cartography. First set of standards published in 1977 and 1979 were the result of earlier work (1974)--CORE--of the Special Interest Group on Graphics (SIGGRAPH) of the Association for Computing Machinery (ACM). From 1979, the ANSI Technical Committee on Computer Graphics used CORE to create two and three dimensional standards based on a graphical kernel (GKS--graphical kernel system). Final report used to generate FIPS #173. Also see FIPS.

answering tone

signal sent by the called modem to the calling modem over public telephone networks--indicates readiness of the called modem to accept data.

Antarctic Circle

parallel of latitude at 66.5 degrees south latitude.

anti-aliasing

removal or smoothing of the staircase effect (aliasing) of computer-generated diagonal lines; also line smoothing.

antihalation

a film backing that absorbs any light rays that penetrate the film base during exposure.   This coating prevents reflection back to the emulsion--a desirable feature.

antireflection coating

single or multilayer coatings applied to a surface to decrease its reflectance and increase its transmission capabilities over a specified range of wavelengths.

APD

see Avalanche Photodiode.

aperture

an opening at the lens mounting of a camera that controls the exposure of a film to light rays.

aperture card

a data processing card to mount a microfilm frame.

API

see Applications Program Interface.

APL

see A Programming Language.

apparatus case

case that holds various equipment such as inductors and repeaters.

APPC

see Advanced Program-to-Program Communications.

application layer

seventh level of ISO's OSI (Open Systems Interconnection) reference model; defines the interface of application software with a network's operating system.

application program

(general)   program to perform a specific user function; (data communications) program used to connect and communicate with sets of terminals to perform functions specific to terminal users.

application software

see application program.

Application Specific Integrated Circuit, ASIC

usually designed by the customer with automated tools provided by the manufacturer.

Application(s) Program(ming) Interface, API

a programming language extension that provides a high-level interface between an operating system and the application programmer.

approach signal

fixed signal used to govern the approach to another signal; if operative, so controlled that its indication furnishes advance information of the indication of the next signal.

APSRS

see Aerial Photography Summary Record System.

arc

common boundary between adjacent mapping units. Also, part of the perimeter between adjacent nodes of a two dimensional closed and bounded figure; an edge joining nodes in a directed graph.

arc data

data that codes the location of linear map features or the borders of polygonal map features.

ARC/INFO system

a vector-based GIS system, produced by ESRI Inc., based on relational data. The ARC/INFO environment consists of a set of parameters used to define a variety of display, editing, and data manipulation conditions that are active until changed by the user.

architecture

the manner in which a computer, network, or program is structured. An architecture is said to be "closed" if it is compatible only with hardware and software from a single vendor. It is said to be "distributed" if it uses a shared communications medium. It is said to be "open" if it is compatible with hardware and software from a variety of vendors.

archive tape

a backup magnetic storage tape used to restore files in the event of accidental data loss; usually stored off-site.

ARCNET

see Attached Resources Computing NETwork.

archival storage

computer peripherals used to store programs and data bases outside the usual addressable memory of the computer.

Arctic Circle

parallel of latitude at 66.5 degrees north latitude.

area

spatial measure of a bounded, continuous, two-dimensional region. The boundary, or portion of the boundary, of the region is not necessarily included in the measure. The problem of measuring area is a difficult one both in terms of the mathematics involved and in terms of real-world considerations. Regular geometric areas and volumes are easy to measure by well-known formulae; however, many real-world areas are not regular geometric figures. Thus, a number of compromises need to be made; a given region can be approximated as a whole or it can be approximated by dissection by a sequence of smaller shapes. Then again, assumptions on the way data is gathered or processed may be made in such a way as to produce regions whose areas are easy to measure. Whether one takes naturally-occurring regions, or structures the shape of the region, the measure is generally a trade off between geometric and real-world accuracy.

Methods for finding areas of regions are of varying mathematical complexity. The simplest are well-known measures such as:

circle: A=pi * radius squared.

square: A= product of two sides.

rectangle: A = product of adjacent sides.

triangle: A= 0.5 times the base times the height--half a rectangle.

n-sided convex polygon:

A=the sum of (n-2) triangular areas formed by joining
one vertex to each of the others.

closed curve formed from curves of known equation:

find the area as a sum of integrals; if the functions describing the curved boundaries are not integrable by elementary means, use approximation techniques for integration (Simpson's or trapezoidal rule or elliptic integrals) or for fitting an equation to the given curve (such as cubic spline interpolation)

Techniques involving the dissection of a large region into smaller ones reduce the error in approximation. These techniques are derivative of material in the mathematical field of "real analysis" (measure and integration) whose theorems often rely on the concept of a limit and the ability to partition areas into infinitely small pieces. Clearly, these assumptions are impossible to realize in the world of digital mapping in which the pixel is the smallest unit; there has been little direct use made of real analysis in the world of digital mapping, although there could be at the research level. From a practical standpoint, the following alternatives for measuring area are often used:

grid cells

a mesh of graph paper is overlain and cells are counted; decisions need to be made along the boundary.

transects

a series of parallel lines contained within the region may be used, their lengths measured, and then added--the sum is then multiplied by the distance between transects times the number of spaces in order to determine the area. From a theoretical standpoint, one might see this as an application of the Fundamental Theorem of Calculus.

point sampling

a lattice may be superimposed and weights assigned to points, consistent with a real-world situation; when added, they give an approximation to area. Pick's Theorem, see affine, is an example of this sort of method. There are many others, some based on lattices with regularly spaced points and some with randomly spaced points (in which case they are not technically lattices).

summary curve

approximation of a given region by a "summary" curve whose area is easy to find as a sum of triangular subregions.

fractal measure

measures the capability of a given curve to fill space. See Mandelbrot, The Fractal Geometry of Nature (Freeman, 1983), and earlier mathematical issues directly related from calculus, such as the relationship between differentiability and continuity.

area symbols

a symbol used consistently across the surface of a map to indicate some common attribute; thus, all deserts might be coded with a dot pattern resembling sand. Shading is also a form of an area symbol that might be used to show relative slope--with steeper surfaces colored darker shades of a pre-selected color.

areal

adjectival form of area; e.g., areal as opposed to temporal unit. Often found in reference to topographic features such as vegetation or soil type.

areal data

two-dimensional geographic data with attributes such as climatic type, national sovereignty, language groupings, or others whose coverage is "wide." Contrast with linear data, such as village distributions in river valleys, whose coverage is "narrow."

Arithmetic Logic Unit, ALU

part of the computer within the CPU that performs the numerical operations on which other programs are based.

arithmetic mean

the sum of the values of the observations divided by the number of observations--the average. The arithmetic mean, and other similar numerical tools, are useful for reducing a large amount of cartographic data to a single value that is easy to map. Thus, lists of data concerning climate or other physical or cultural characteristics of a region may be compressed to a small number of values that can be distinguished by shading on a map.

arithmetic operators
(GIS) arithmetic operations (addition, subtraction, multiplication, and division) performed on cell maps.

ARPANET
see Advanced Research Projects Agency NETwork.

ARQ
see Automatic ReQuest for retransmission.

array
a method of storing data in which the entries are grouped together for reference purposes. A one-dimensional array A, a vector, consists of entries A(I), for a range of values of I, usually from 1 to some value N. A two-dimensional array, or matrix, consists of entries A(I,J), in which one normally thinks of a typical entry as being in the Ith row and the Jth column. For example, pixels on a computer screen can be addressed in this way. By analogy, arrays of any dimension can be defined by specifying more subscripts. For example, if each layer in a GIS is a matrix, then A(7,I,J) might refer to the entry in row I, column J of the 7th layer.

array processor
a computer that operates in parallel (simultaneously) on many elements of an array. Also, a math coprocessor designed for fast calculations, through parallel operation on various elements, with arrays of numbers; usually incapable of other kinds of operations. Often used in image processing and in signal analysis.

Artificial Intelligence, AI
computer simulation of inductive reasoning based on a set of logical rules. Artificial intelligence programs are also referred to as expert systems. The term comes from the fact that these programs are constructed so that they reject previous possibilities which failed when similar situations were encountered earlier. Thus they "learn" from experience, hence simulating intelligence. See also neural networks.

artwork
graphic materials prepared for reproduction by any of a variety of means. These are often prepared in a digital format suitable for reproducing on a printer or a plotter.

ASCII
see American Standard Code for Information Interchange.

ASCII file
a text-only file.

ASCII terminal
asynchronous terminal; dumb terminal--terminal that uses ASCII.

ASD
see Auxiliary Storage Device.

ASIC

see Application Specific Integrated Circuit.

aspect

horizontal direction that tells in which direction a surface slope faces--measured in degrees (and fractions thereof) clockwise from north. An east-facing slope has aspect of 90 degrees; a south-facing slope 180 degrees; and, a west-facing slope 270 degrees. Generally, it is the compass direction measured from magnetic north of the line of steepest slope at some pre-selected location.

aspect ratio

computer graphics--ratio of the horizontal dimension of an image to its vertical dimension. Varying this ratio causes distortion. Thus, in high definition television (HDTV), for example, one needs not only to worry about increasing the number of horizontal and vertical lines, but also their relation to each other--the aspect ratio. Standard square pixel space has an aspect ratio of 1:1 which is well-suited to image processing; broadcast video has a ratio of 4:3 which should be corrected to give a true picture coming from any digitization process of these images.

ASR

see Automatic Send/Receive teleprinter.

assembler

a computer program for translating instructions from a lower level computer language (source language) to machine language, binary instructions.

assembly language

a low-level machine oriented computer programming language in which individual statements represent discrete machine operations.

assigned pairs

the total number of pairs assigned for customer service, whether working or idle assigned.

assignment office

the telephone company office that maintains records of the status of all cable pairs and assigns an available pair for connection to a customer's premises in response to a customer's request.

associated data

see attribute.

Association of American Geographers, AAG

U.S. geographic organization (founded 1904) that holds regular annual, and more frequent regional, meetings for its members. One of its Specialty Groups is devoted to Geographic Information Systems; another to Quantitative Methods and Mathematical Models; and yet another to Microcomputers.

assumed plane coordinates
a plane coordinate system with assumed reference axes, often established for the convenience of a surveyor.

asynchronous character
binary character used in asynchronous transmission. In format, it contains equal-length bits, one of which is a start bit defining the beginning of the character, and one or more of which are stop bits defining the end of the character.

asynchronous communications
see asynchronous transmission.

asynchronous modem
a modem that uses asynchronous transmission.

asynchronous terminal
see ASCII terminal.

Asynchronous Time-Division Multiplexor, ATDM
a time-division multiplexor that multiplexes asynchronous signals by oversampling them.

Asynchronous Transfer Mode, ATM
a type of packet switching that transmits fixed-length units of data; it is asynchronous in that the recurrence of cells does not depend on the bit rate of the transmission system - only on the source requirements.

asynchronous transmission
transmission of data via serial lines without any specific timing pattern. Characters are sent one bit at a time and the spacing between bits need not be of equal length. Asynchronous characters are used, in which extra bits are added to the beginning and end of each data character (start and stop bits) to synchronize the sending and receiving devices; commonly used with modems on PCs and minicomputers. No additional timing information need be sent because the asynchronous characters handle timing as a consequence of their format.

ATDM
see Asynchronous Time-Division Multiplexor.

atlas
a logically ordered collection of maps, often bound together as a book.

Atlas GIS
a geographic information system for the PC developed by Strategic Mapping; non-topological vector format. See references in front matter.

ATM
see Asynchronous Transfer Mode.

atmospheric window
   radiation wavelength ranges of values--within a range, radiation with wavelength falling in the range can pass through the atmosphere with low attenuation.

Attached Resources Computing Network, ARCNET
   popular network architecture that uses the token-passing bus protocol.

attenuation
   the decrease in the flux density of a parallel beam of energy with increasing distance from the energy source, as for example, the decrease .in the strength of a signal as it passes through a control system or transmission medium. This decrease is generally the result of absorption, reflection, diffusion, scattering, deflection, or dispersion rather than a geometric consequence of spreading as the inverse square of distance. In telecommunications, attenuation increases with signal frequency and cable length; measurement in the telecommunications environment is in deciBels. In particular, optical fibers can be classified as high-loss (over 100 dB/km), medium loss (20-100 dB/km), and low-loss (less than 20 dB/km).

attitude
   the angular orientation of a camera, or of the photograph taken with that camera, with respect to some external reference system; usually expressed as tilt, swing, and azimuth or roll, pitch, and yaw. In a satellite system, the angular orientation of a remote sensing system with respect to an external (to the camera) geographic reference system. Plate III shows an oblique view of San Francisco Bay.

attribute
   a distinct characteristic of an object or entity; analogous to a record field. On a map, data about map features that are stored as non-graphic, often alphanumeric, text in a database format as a record. Attribute data about telephone lines (or other utilities) might list the number of the pole, its height, and the material it's made of (see Plate V). Attribute data about a city map might include home ownership information, land use type, or zoning information.

attribute data
   (GIS) - a set of (usually) alphanumeric data describing characteristics of real-world entities or conditions. A small quantity of it might describe a polygon label; large quantities (such as census data) are often maintained as separate data sets that can be related to a map by names or codes.

attribute input
   a data input function.

attribute query
process to select data items from a file system.
audio graphics
transmission of graphic and textual content over narrowband telecommunications channels such as a telephone line or radio subcarrier.
audio teleconferencing
interactive electronic voice telecommunications between two or more groups, or more than two individuals, in separated locations.
audit data/table
a table of information describing the content of a map--items, perimeters, and areas.
auto-answer
modem used on public telephone network that automatically answers incoming calls with an answering tone.
auto-dial
modem used on the public telephone network that automatically dials numbers.
autocarto
American Cartographic Association congresses focused on computer-assisted cartography and GIS.
autocorrelation
in statistics, the correlation of an ordered sequence of observations with the same sequence of observations; the lag is the size of a stated interval offsetting the second sequence from the first. Often used with reference to spatial data; the degree to which the values of an attribute of two distinct objects covary in relation to the distance between them. Spatial autocorrelation can find the best compromise line between overlapping images. Also autocovariance.
autocovariance
see autocorrelation.
automated cartography
computer assisted mapping system using any of a variety of software on a personal computer, a microcomputer, a minicomputer, or a mainframe. Characteristically, there are three elements to such a system: hardware, software, and database. The graphics interact dynamically with the data, so that an update in the data set is automatically reflected in the graphic display and analysis of that data set.
automated contouring
see contour.
Automated Geographic Information System, AGIS
another word for GIS, using a digital computer.

Automated Mapping system, AM
   computerized system used to draft and generate maps; a graphic tool which may or may not also be capable of geographical analysis.

Automated Mapping/Facilities Management, AM/FM
   AM/FM - process that translates cartographic information into a digital format stored for future manipulation. More specifically, an automated mapping system designed to record spatial data about a variety of aspects of real-world facilities, such as sewer or power lines, water mains, or other public works and utility applications. The spatial data involved usually does not have the degree of detail needed for property line determination and other surveying issues. These systems process both graphic and non-graphic data and can manage geographically distributed facilities, overlay combinations of features, analyze flows in networks, and define geographic regions to fit *a priori* criteria (Plate V). This Plate shows a typical digital map of facilities mapping.

Automatic Baud Rate detection, autobaud, ABR
   (computer terminology)   Process in which a receiving device controls the code level, speed, and stop bits of an incoming data stream by examining the first character of the stream. Often, the first character is a pre-selected sign-on character. Using ABR, the receiving device can accept data from a number of different transmitting devices operating at different data transmission rates. With ABR, there is no need for advance configuration of the receiving device to correspond with each of the different data rates.

automatic digitizers
   see digitizer.

Automatic ReQuest for retransmission, ARQ
   error control strategy; the receiver informs the transmitter which transmission blocks were received; the transmitter then re-transmits any blocks that were not successfully transmitted in the first place.

Automatic Send/Receive teleprinter, ASR
   teleprinter equipped with tape (paper or magnetic) or a solid state buffer allowing it to receive and to transmit data in an unattended mode.

automatic shut-off valve
   protects the manifold assembly and cables from a pressure loss if a failure occurs in the supply line of a cable pressurization system.

Auxiliary Storage Device, ASD
   mass storage devices such as disk and tape drives. These devices maintain digital files that are directly accessible to a processing unit.

availability

a performance measure of equipment, systems, or networks. Availability equals Mean-Time-Between-Failures divided by the sum of Mean-Time-Between-Failures and Mean-Time-To-Repair.

available pairs

total pairs in an entity (such as a pair group, EFRAP section, CO, complement, serving area, or allocation area) including working, spare, and defective pairs. The pairs may reside in more than one cable.

avalanche multiplication

in a semiconductor, the rapid increase in the number of carriers for conduction, caused when the semiconductor is subjected to high electric fields. The initial rapid increase causes still further increase. These multiplicative effects are employed in avalanche photodiodes.

Avalanche PhotoDiode, APD

a photodiode designed to take advantage of the avalanche multiplication of photocurrent phenomenon for fiber optic signal transmission.

average

a numerical measure of central tendency. There are many kinds of averages; in common usage, "the average" usually refers to the arithmetic mean.

AVHRR

see Advanced Very High Resolution Radiometer.

AVIRIS

see Advanced Visible/Infrared Imaging Spectrometer.

azimuth

in its simplest expression, horizontal direction measured in degrees clockwise from north. Due east has azimuth of 90 degrees. Azimuth is a more abstract term than aspect, which refers to slopes. The calculation of arbitrary azimuths relative to the geographical coordinate system is numerically complicated and quite specialized.

azimuthal projection

projection of the earth-sphere into a plane tangent to the sphere at any point of the sphere. The condition of projection means that there is no deformation at the center of the projected map, that great circles through the point of tangency are projected to radial lines through the map center, and that the projected map is centrally symmetric about the center of the map projection--scale varies outward from the map center (linearly, angularly, or areally). Direction measured from the central point of the projection is true on azimuthal projections. There are many categories of azimuthal projection. Some of the more common ones are listed below.

Gnomonic projection: center of projection is center of the earth sphere; therefore, less than half a sphere can be shown on a map. On this projection, all great circles are shown as straight lines: direction is true. Thus, it has some use for navigation, but deformation increases rapidly away from the map center.

Azimuthal equidistant projection: mathematically constructed; linear scale does not vary along the azimuths radiating outward from the center, so that direction and distance are accurate when measured between points on the same radial passing through the center (but not between points not on the same radial). A very useful projection when distance and direction from a central point are important. In Figure 9, the azimuthal equidistant projection is centered on the North Pole. In Figure 10, it is centered on the South Pole.

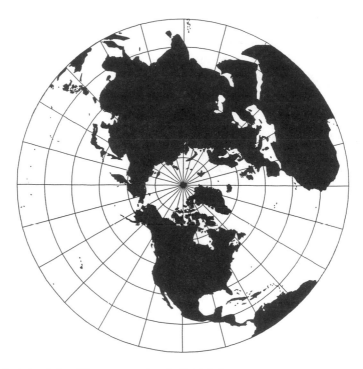

Figure 9. Azimuthal equidistant centered on the North Pole.

Notice that in these equidistant projections the even spacing between the parallels of latitude suggests that distance along a radial is true. This fact causes the landmasses near the perimeter, such as South America in Figure 10, to appear unusually large. When relative areas, rather than relative distances measured outward from the center, are important, one might therefore choose an equal area projection. One that is azimuthal is Lambert's azimuthal equal area: Figure 11 shows one centered on the North Pole and Figure 12 shows one centered on the South Pole. The stretching of the areal size, as one approaches the map perimeter of the Lambert projections, is controlled as suggested in the spacing between parallels in the underlying graticule (they are closer together as one moves away from the projection center). In the Lambert projection, shape is sacrificed in order to make areas comparable.

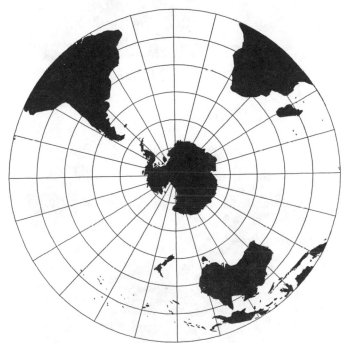

Figure 10. Azimuthal equidistant centered on the South Pole.

Orthographic projection: center of projection is at infinity, so that an entire hemisphere, that looks like a globe, is easily portrayed. Deformation is large, but often not apparent because of the globe-like appearance. Useful for attractive visual displays, especially when sphericity is important; not good if accuracy is significant.

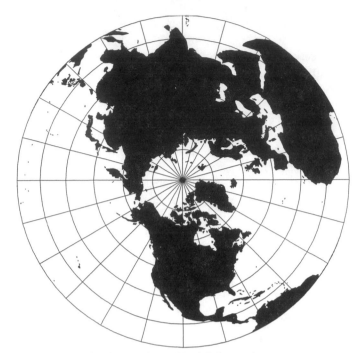

Figure 11. Lambert Azimuthal Equal Area (North Pole center).

Figure 12. Lambert Azimuthal Equal Area (South Pole center).

AZP
   see Algemeine Zeichen Program.

B b

B-spline
   Bi-cubic spline. A curve formed from the sum of other spline curves that allows local curve fitting in a simple manner. Useful for smoothing curves generated from empirical data; useful to provide a fit to a bounded set of points, rather than for extrapolation outside a bounded interval.

back board
   mounting boards, where communication system connecting blocks are placed to terminate cable.

back-up file
   a duplicate file created to prevent data loss in the event the original file is damaged or lost.

backbone
   (in packet-switched networks) a network designed to interconnect lower-speed channels; the major transmission path for a Packet Data Network.

backcloth
   see backdrop.

backdrop
   the "stage" behind the "actors" on a map--the backing on which mapped features provide information on location (relative position) of the primary subject matter of the map. In contrast to background, which is a more general term, backdrop is more concrete and visual. In Plate II the backdrop on the left side suggests the terrain backdrop of the urban subdivision (upper right); in Plates VII and VIII, the road network is the backdrop for school allocation. See base map.

background (processing)
   in a computer with multitasking capabilities, the environment in which activities such as printing are executed while the user works with an application in the foreground. With background printing (for example) the user can continue to work as printing takes place. Generally, background refers to something behind the primary task or subject.

backlight
   as a verb, to pass light through copy from behind the image.

backlit display
   in a computing environment, often a liquid crystal display (LCD).

backplane

the parallel circuit path that carries data and address information throughout the computer system.

backscatter

in reference to radiant energy, its scattering back toward the source of the incoming ray.

band

a range of spectral values: wavelengths, radar (or other) frequencies, energy absorption, tracks on a magnetic drum; also called channel.

Band Interleaved by Line, BIL

values for each line in a raster are stored sequentially in the database prior to going to the next line.

Band Interleaved by Pixel, BIP

values for each pixel in a raster are stored in the database prior to going to the next pixe.

Band SeQuential, BSQ

data array for each variable is stored separately and is not dependent on any other variable.

band-pass filter

a filter on a wave that permits only a specified band of wavelengths to pass and blocks all others.

bandsplitter

multiplexor dividing the bandwidth into independent narrower bandwidth channels, each suited to transmitting data at a rate less than that of the composite channel rate.

bandwidth

term expressing transmission speed of data and relative range of frequencies that can carry a signal without distortion on a transmission medium. The higher the value, the sharper the image. The capacity of an optical fiber (or other medium) for the transmission of information, usually expressed in bits of information transmitted in one second for a specific length of optical waveguide. The phase "10 megabits/sec/Km" or "bandwidth of 10" means that the line can transfer data at the rate of 10 million bits per second. Optical fiber bandwidth is limited by pulse spreading or broadening due to dispersion; the broader the bandwidth the greater the message carrying capacity of a medium. It also describes the range of frequencies available for signaling, with the difference between the highest and lowest frequencies in the range measured in Hz.

barriers

static or dynamic factors which block transmission of information or flows of goods and services. Typically, rivers, time, weather, and a host of other terrestrial conditions and electronic conditions, can serve as barriers to distance or connectivity analyses. The concept of barrier is a fundamental one in geographical analysis; it is also a dynamic one. What is a barrier to some, may be a conduit to others. The river that needs to be bridged to move from one country to another may also serve as a conduit for freighters bringing needs raw materials to local factories.

base data

basic level of map data used as a foundation on which to place other data for geographic comparison.

base line

surveyed reference line upon which other surveys are based.

base map

an accurately scaled map that shows certain physical and topographical features, such as highways, streets, railroads, rivers, and lakes. When this map, containing reference information, is used as a foundation on which to build other maps by superimposing attribute data, it serves as a base map. Base maps and the layers superimposed on them are often used for purposes of geographical comparison and correlation. Base maps generally contain data, such as river positions and coastal boundaries, that are slow to change. In that regard, the left half of Plate II might serve as a base map--in contrast to Plate V; although neither is the sort of outline map of continental boundaries that many would associate with the word base map (as for example in the Goode's base map series of the University of Chicago). On Plate IV, for example, a base map is critical because the data alone, displayed as a collection of disks, makes no sense without the base map which provides locational information (orientation). See backdrop.

base memory

the part of a computer's Random Access Memory (RAM) devoted for user programs.

base station

central radio transmitter and receiver that maintains communication with mobile radiotelephones within a geographic cell.

baseband

in a digital data system, baseband refers to a digital stream that carries just one information channel. LAN terminology--a communications technique in which the signal is put directly on the cable, without modulation, in digital form.

baseband modem

line driver or local dataset.

BASIC

Beginner's All-purpose Symbolic Instruction Code - a high level programming language similar in structure to FORTRAN but easier to learn (although widely criticized--programs execute slowly). Interpreted rather than compiled.

basic cover, (aerial photography)
photographic coverage recorded in such a way that later photographs of the same area will reveal topographic (or other) changes. Plate II-- lower right half suggest basic cover of the region.

Basic Exchange Telecommunications Radio Service, BETRS
radio transmission system used by telephone companies to provide basic service in remote areas, especially where terrain presents difficulties resulting in extra expense (or lack of practicality).

basic input-output system, BIOS
programs encoded in ROM in PCs which transfer data and control instructions between the computer and peripherals. A basic set of operating instructions used by many microcomputers for internal and external communications.

Basic Rate Interface, BRI
ISDN transmission with 2 "B" (Bearer) channels at 64kbps and 1 "D" (signaling) channel at 16 kbps ("zB+D"); see PRI.

Basic Telecommunications Access Method, BTAM
IBM software routing--basic access method for 3270 data communications control.

batch processing
the sequential manipulation of large groups of data; a mode of processing in which commands are submitted for subsequent execution rather than processed interactively; input data is prepared off-line and processed in batches. During program execution, the user is not in direct communication with the processing unit. The user creates a job, containing data that includes information required to execute the program, the destination of the output, and the identification of the user (often by coded identification). Synonym: off-line processing.

batch queue
an execution queue controlling batch jobs.

baud rate
originally the unit of speed of transmission of information by telegraph signaling such as Morse code; now the baud is defined as the reciprocal of the length in seconds of the shortest element of code in a transmission. It is sometimes used as a synonym for bits-per-second (bps). Special modulations, however, can deliver a bits-per-second rate greater than the baud rate. Thus, a more general conception of baud rate is as a measure of signaling speed, as one signal event per second, describing the average frequency (modulation rate) of the signal.

Baudot code
in contrast with ASCII and EBCDIC code. This code is used primarily in teleprinter systems that add one start bit and 1.5 stop bits. In this asynchronous transmission of data, 5 bits represent a single character. Named after Emile Baudot.

bayonet Neill-Concelman, BNC
bayonet-locking connector for miniature coaxial cable, as contrasted with a threaded Neill-Concelman connector.

B-CDMA
see Broadband Code Division Multiple Access.

bearing
horizontal (east-west) angle of a directional line measured as suggested in the following examples--northeast has a bearing of 45 degrees east of north; southwest has a bearing 45 degrees west of south.

Behrmann cylindrical equal area projection
relative area measurements are accurate; shape is not true. Sometimes referred to as a "Peters" projection. See cylindrical projections for a figure and further discussion.

Bel
unit of measure equal to 10 deciBels.

Bell 43401
Bell Publication. Defines requirements for transmission over telco-supplied circuitry with dc continuity (are metallic).

Bell modems
by number, these AT&T modems (and compatibles) are:
103 - an AT&T 0-300 bps modem providing asynchronous transmission with originate/answer capability--also describes a modem compatible with this one.
113 - like a Bell 103, but with originate or answer (but not both) capability.
201 - 2400 bps modem providing synchronous transmission.
201B - designed for leased line applications.
201C - designed for public telephone network applications.
202 - 1800 bps modem providing asynchronous transmission requiring a 4-wire circuit for full-duplex operation. Also, a 1200 bps modem providing asynchronous transmission over 2-wire, full-duplex, leased line or public telephone network applications.
208 - 4800 bps modem providing synchronous transmission.
208A - designed for leased line applications.
208B - designed for public telephone network applications.
212 and 212A - 1200 bps modem providing asynchronous transmission or asynchronous transmission for use on the public telephone network, over full-duplex lines.

Bell System Practice, BSP

provides material describing procedures, safety requirements, standard symbols, and so forth.

benchmark (bench mark)

a standard measure of some aspect of computer performance, typically tests conducted prior to purchase; common benchmarks include measures of integer operations (Fibonacci series, Sieve of Eratosthenes), system operations (Dhrystone), the Gibson mix of tasks (MIPS), and floating point operations (Livermore Loop, Linpack, Whetstone, Quicksort). Also, a vertical control point above or below a base level marked by a monument. A benchmark test evaluates the capabilities of a computer with respect to the customer's needs.

BER

see Bit Error Rate

Bernoulli box

removable mass storage system for PCs and Macintosh computers developed by Iomega Corporation.

BERT

see Bit Error Rate Test

best linear unbiased estimate

value calculated from an interpolation function.

beta test

test of hardware or software performed in a normal operating environment, as contrasted with alpha test, conducted in the developer's environment.

BETRS

see Basic Exchange Telecommunications Radio Service.

Bézier curve

(computer graphics) non-uniform curves drawn by bending a line segment at strategic locations using its distinguished graphic "handles," usually at four corners of a bounding graphic box. Can be made to fit continuously with others of its kind.

bidirectional coupler

a coupling device for optical fiber circuits that allows signals to propagate through the fibers in both directions.

BIL

see Band Interleaved by Line.

bilinear interpolation

used to resample a rasterized object in order to create a new rasterized object.

binary

in digital mapping terminology, often used to mean having two possible states or values (e.g., "on" and "off"). Since binary arithmetic uses only the numerals 0 and 1, it is ideally suited to computer use, where information is stored by means of objects which can have only two states. The algebraic manipulations are identical in structure to those for the decimal system. Examples to illustrate the mechanics of decimal and binary arithmetic are given below.

The decimal expansion

$$23 = 2*10^1 + 3*10^0$$

has as its binary counterpart

$$10111 = 1*2^4 + 0*2^3 + 1*2^2 + 1*2^1 + 1*2^0$$

To add binary numbers, it is possible to convert them to decimal numbers, and the decimal numbers and convert back to binary. It is generally easier, however, simply to add them as binary numbers using the same principles as for addition in the decimal number system. The key is to remember to carry--thus, 2+2=4 does not require any carrying as a decimal sum; as a binary sum it does--$1*2^1 + 1*2^1 = (1+1)*2^1 = 1*2^2 + 0*2^1 + 0*2^0 = 100$.

binary code

a code, consisting of strings of 0's and 1's, used to communicate with a digital computer; based on a number system with a base or radix of 2. As contrasted with octal, decimal, and hexadecimal.

BInary digiT, BIT

one unit of information in binary notation. Contraction of binary digit - the basic element of a binary code, represented by a pulse (binary value 1) or no pulse (binary value 0) or some comparable two-valued encoding. See also byte, word.

binary encoding

transforming discrete levels into a two-state code for processing by digital techniques.

binary file

file written using binary characters that are in machine-readable form.

binary raster

raster with cells coning only the values of 0 and 1.

binary search

computer search algorithm that starts in the middle of a sorted database and checks above and below the midpoint, eliminating half the possibilities each time. Faster than sequential sorting.

binary state

either of the two conditions of a bistable device, the "one" state or the "zero" state.

Binary Synchronous Control, BSC

a character-oriented protocol that uses a standard set of transmission control characters for synchronous transmission of binary-coded data between computer systems. It was originally developed by IBM. Data transmission is in 8-bit strings each representing a portion of the message. In wide use in the 1970s and early 1980s, this protocol was not well-suited to complex network interactive processing of data; hence, bit-oriented processing, better suited to these needs, was developed in the late 1970s.

Binary Universal Format for data Records, BUFR

a weather service X.25-based protocol.

binary word

(telecommunications) a combination of 8 pulses or no-pulses corresponding to the PAM sample of one channel; a binary word requires eight time slots and is 5.2 microseconds long. See also word.

Binary-Coded Decimal, BCD

a digital system that uses binary code to represent decimal digits. Using this code, each character (number, letter, or punctuation mark) can be represented in a computer using an 8-bit byte.

binder group

a group of 25 pairs of wires in a cable. The group is wound with a colored tape that indicates its numeric position in the total count of pairs in the cable. Used by cable splicers to identify correct pairs for splicing.

BIOS

see Basic Input-Output System.

BIP

see Band Interleaved by Pixel.

BiPolar Non-Return-to-Zero, BPNRZ

BiPolar Non-Return-to-Zero, same as BPRZ except transitions between adjacent 1's do not stop at zero level.

BiPolar Return-to-Zero, BPRZ

BiPolar Return-to-Zero, a three-level code in which alternate 1's change in sign; for example, 1011 becomes +1, 0, -1, +1) and transitions between adjacent 1's pause at the zero voltage level.

bipolar transmission, polar transmission

digital transmission of binary data in which negative and positive states alternate.

BiPolar Violation, BPV

a violation of the alternating +1/-1 pattern in a ternary (3 level) code.

BIT

see BInary digiT.

Bit Error Rate, BER

the ratio of erroneous bits to the total number of bits in a transmission.

Bit Error Rate Test, BERT
a means of determining the bit error rate.

bit plane
part of a data structure of superimposed grids of cells of values 0 or 1.

bit rate
a measure of the number of bits (binary digits) transmitted in a given time interval; the time interval is usually one second so that a bit rate ordinarily is stated in bits per second.

bit-oriented
in contrast to character-oriented, describes communications protocols in which information may be encoded in fields as small as a single bit.

bit-stuffing
in Synchronous Data Link Control (SDLC), the inclusion of a binary 0 bit in a data stream to avoid confusing characters that might otherwise be mistaken for one another. Generally, the inserted zero is removed by the recipient. See also: zero insertion.

bitmapped font
text font defined using bitmapped images; these fonts look good when printed at the scale at which they are drawn. When enlarged, there may be considerable aliasing.

bitmaps
(graphics) in a video image stored in a computer's memory, each pixel is represented by bits. PC generated graphics from "paintbrushes" are often described and stored as bitmaps--each sampled point is described in two dimensional space according to its color and intensity. Bitmaps use large amounts of RAM.

BITNET
a wide area network linking more than 1000 universities. The acronym stands for "Because It's Time NET"work.

Bits Per Inch, BPI
a measure of the density of information on a magnetic tape; 800 bpi, 1600 bpi, 6250 bpi, and 6500 bpi are common recording densities.

Bits per pixel
the number of data bits represented by each pixel; also, pixel depth.

Bits Per Second, BPS
a measure of digital data transmission speed, or data rate. When combined with metric prefixes, such as kbps, these words measure the number of bits per second (thousand bits per second in this example).

blackbody

theoretical matter that changes heat energy, at the maximum rate allowed by thermodynamic laws, into radiant energy. Also, a corrective target used to calibrate ground based sensors and satellite sensors.

blacklight

an ultraviolet flourescent lamp, treated with phosphors, which when used as a light source for photographic exposure, screens out almost all visible light.

block

a unit of measurement for the space in a computer's memory consumed by a file; a group of records treated as a logical unit of information. Also, transmission block.

Block Check Character, BCC

error detection character added to the end of a transmission.

block diagram

a perspective drawing of a cross section of the earth's surface--a "block" cut from the earth's surface, showing a profile across the land surface. Often used to show geological features and landforms on the earth's surface. A good rendition is highly graphic and looks quite realistic to most viewers. Block diagrams have long been in use as graphic displays. See Plate IX.

block i/o

a data accessing technique in which the program manipulates the blocks (physical records) that make up a file, instead of its logical records.

block transfer

high speed transfer between disk files and memory that bypasses the CPU.

block-multiplexor channel, byte interleaved channel

(IBM systems)   A multiplexor channel that interleaves bytes of data, as contrasted with a selector channel.

blocking

(LAN terminology)   Inability of a PABX to service connection requests.

blue-line

paper copy of original plat either black and white or blue and white.

BNC

see Bayonet Neill-Concelman.

board

electronic circuit board; used in a microcomputer to install added hardware.

board digitizing
   interactively transforming analog vector data into digital vector data using a cursor, menu, tablet, and graphics terminal or work station; also known as trace digitizing.

BOC
   see Build-Out Capacitor.

BOL
   see Build-Out Lattice.

bond
   a metal braid placed across the metal portion of the sheath while making a splice to provide electrical continuity of the cable sheath.

Bonne projection
   equal area projection with a standard central meridian. Deformation increases directly with distance from the central meridian. Thus, the Bonne projection is better for maps of greater north-south than east-west extent. See equal area projection for further general discussion.

Boolean algebra
   the mathematics of logic which uses alphabetic symbols to represent logical variables and "1" and "0" to represent states; there are three basic logic operations to this algebra; AND, OR, and NOT; named after George Boole (1815-1864), English mathematician.

Boolean operations
   operations of union, intersection, complement, and exclusion, used to link logical combinations of data.

boot
   short for "bootstrap", a procedure by which a computer "pulls itself up by the bootstraps" in order to become fully operational (i.e., the loading of auto-starting programs from external storage).

border matching
   the process of joining together individual digital data sets utilizing computer application software.

boundary
   a line, straight or curved, separating contiguous regions of land, water, air, or other parcels. Boundaries may be formed on political or cultural bases as well as on many others. Boundary is an abstract concept that is seldom possible to make "accurate" on any map. Because boundaries are difficult to make precise, they have historically been the source of many disputes. Boundaries also produce specialized effects in the zone near them--they are often more difficult to cross than land that is not directly adjacent, and the result is a change in the land use pattern near the boundary. The concept of boundary is one of primary importance in geography and in the mapping of geographic regions. All too often, its importance is underestimated or even

overlooked. In the absence of an actual boundary line, the red lines (streams) in Plate VI suggest the boundary of the Mississippi River drainage basin.

boundary line

line marking the boundary between contiguous geographical regions.

boundary monument

physical construct marking a corner location on a boundary.

boundary survey

survey to establish a terrestrial boundary line to use as a boundary line on a map. Also, cadastral survey or land survey.

bounds

see, metes and bounds.

boxcar classification

automated image classification using three simple data ranges, into which to sort three coregistered images, such as red, green, and blue. The three ranges form a "box" in three dimensional space; hence, the name.

BPI

see Bits Per Inch

BPNRZ (or Bipolar NRZ)

see BiPolar Non-Return-to-Zero.

BPRZ (or Bipolar RZ)

see BiPolar Return-to-Zero.

BPS

see Bits Per Second.

BPV

see BiPolar Violation.

break

an interruption or break of one character time (often 110 milliseconds) often used by a receiving terminal to request disconnection, to terminate computer output, or to break the sending device's data transmission.

breakout box

a device for testing of electrical circuits in a cable or a connector.

BRI

see Basic Rate Interface.

bridge [1]

physical device with software that links one network and its resources with another via the OSI data link layer. Contrast with gateway.

bridge [2] or branch splice

the connecting of three or more pairs of conductors in three or more cable sheaths.

bridged tap

an extension of a cable pair beyond the point where it is used, or a branch cable that has bridged pairs. A bridged tap impairs transmission.

brightness

the perception of a color value that reveals differences in luminance. On a gray scale, white has maximum brightness value and black the minimum. Brightness can be computed from LANDSAT and other images using a number of different transformations. Contrast with other dimensions of color of "hue" and "intensity."

broadband

communications channel of bandwidth greater than a voice channel; capable of higher transmission rates than a voice channel; can support ISDN or other integrated voice, data, and video applications.

Broadband Code Division Multiple Access, B-CDMA

in cellular phone and personal communications service, this is CDMA with improved multipath resolution coming as a result of using broadband, rather than narrowband, transmission.

broadband ISDN

broadband Integrated Services Digital Network.

broadband ISDN standardization

some of the standards issues involve user-network interfaces, congestion control, cell formats, services, and operations and management.

broadband network

a communications network that functions with uniform efficiency over a wide band of frequencies. This sort of network has the capability to transmit multiple channels simultaneously and is of great promise in application--broadband graphics communication; broadband medical communication; broadband medical imaging; broadband medical network evolution; broadband movie distribution; broadband packet switching; broadband videotelephony; broadband videotext.

broadcast

transmission of message intended for general reception.

browse

simple search of a database to answer locational queries--including zooming and panning of a map; level of access to a library that permits examining it by not modifying it.

BSC

see Binary Synchronous Control.

BSP

see Bell System Practice.

BSQ

see Band SeQuential.

BTAM

see Basic Telecommunications Access Method.

bubble memory

a type of non-volatile computer memory that uses magnetic domains (bubbles) for data storage.

buffer [1]

an intermediate storage register or device used to hold information temporarily.

buffer [2]

a protective material which is used in cabling optical fiber to cover and protect the fiber.

buffer pairs

in a serving area interface contiguous spare feeder pairs that are flagged as a warning impending feeder pair exhaustion.

buffer zone

a zone of specified distance around map features; both constant-width and variable-width buffers can be generated. Plate X shows 10 mile buffers around South Carolina Interstate highways.

BUFR

see Binary Universal Format for data Records.

bug

programming error that causes a problem in a computer program. One explanation for the word (perhaps apocryphal) is that an early computer program failed to work because one of the vacuum tubes failed because there was (literally) a bug on it.

Build-Out Capacitor, BOC

a capacitor added to a cable pair to correct the pair's electrical length and thus eliminate any impedance irregularity.

Build-Out Lattice, BOL

a build-out lattice is used in the same way as a build-out capacitor.

building cable

cable attached to or placed in a building or between buildings in a conduit, pipe, or tunnel. See Plate V; typical of maps on which this might be displayed.

building nodes

adding nodes in a vector object from data base files or other imported files that contain coordinatized entries.

bulk material absorption

lightwave power absorption occurring per unit volume in the basic materials used to form an optical fiber--generally expressed in decibels/kilometer.

bulk material scattering
   lightwave power scattered per unit volume of the basic material forming an optical fiber. Scattering losses are generally expressed in decibels/kilometer.

buried cable or wire
   the type of construction in which multipair cables or wires are buried in the ground without a conduit.

buried construction
   the type of construction in which multipair cables or wires are buried in the ground without a conduit.

burst mode
   an operating mode in which large continuous blocks of data are transferred, without interruption, between system memory and a peripheral device such as a data acquisition board.

bus, buss
   computer communication lines, generally computer system specific. Information transfer depends on the width (number of bits that can be simultaneously transferred) of these parallel data paths. Mainframe computers often have buses that are 24, 32, 48, or 64 bits wide; mini- or micro-computer buses are generally 8, 16, or 32 bits wide. In LAN terminology--a linear network topology.

byte
   a set of adjacent binary digits stored and retrieved as a unit within a computer. Bytes are often grouped together into words. The most common word lengths are 16 bits (2 bytes)--common in PCs and 32 bits (4 bytes)--standard in mainframe computers and now occurring PCs as well. The capacity of storage devices is frequently given in bytes.

byte stuffing
   in digital data systems, the technique by which a six bit or seven bit word is converted to an eight bit byte for transmission.

C c

C
   capacitor, capacitance, capacity, carbon, coulomb, centigrade, (centi-), Celsius, transistor collector, candle, or velocity of light in air (or a vacuum). Computer programming language.

C connector
   a bayonet connector for joining coaxial cable; the C is for Carl Concelman.

"C" field code
   stands for Construction, build, install, and place.

C language
computer programming language in which much applications software for a UNIX environment is written. C is a third-generation language developed at Bell Laboratories.

C++ language
an extension of the C language useful for object oriented programming applications.

cable
two or more pairs of insulated wires wrapped together with a protective covering, used to connect stations.

Types:

Underground Cable (UG) - Cable placed in conduit and manholes.

Aerial Cable (AER) - Cable strung on poles.

Buried Cable (B, Bur) - Cable placed directly in the ground.

Submarine Cable (SUBM) - cable placed in water.

Trunk Cable (TRK) - Cable between C.O.s

Toll Cable (T) - Cable between C.O.s which involves long distance charges

Building Cable (BLDG) - Cable placed in buildings

in conduit, pipe, or tunnels.

cable marker
a surface flag or pin marking cable location underground. Plate V displays the sort of map on which this object might be noted.

cable number
the unique alphanumeric identifier of each administrative or central office cable in a wire center. Sometimes referred to as cable ID.

cable pair
one pair of insulated wires, within a cable, twisted together, which has its own unique pair number. Loose twisting of the wires helps to cancel out any noise in balanced electrical circuitry.

cable system
in a Local Area Network, the physical medium used to join stations on the LAN; sometimes referred to as the premises network.

cable television services
public and commercial television carried over local and national cable networks.

cable transfer
a subscriber line rearrangement in which a cable pair is severed and the field end is respliced to a different pair located toward the central office.

cable vault

a room in a Central Office (CO) that allows an interface between the outside plant cable network and cables leading to the main distributing frame and central office equipment.

cable-buffering

a process of maintaining air pressure on cable plant. Cable buffering prevents the entry of moisture for the activation of pressure sensor alarms when a sheath or a splice opening is made on pressurized cable.

cable-cap

a cable cap indicates the end of a cable has been sealed with a "cap."

cable-count

identifies the individually numbered cable pairs contained in a particular section of cable. For example, 61, 701-900.

cable/pair range

a combination number consisting of a cable number plus two numbers representing the starting and ending pair numbers of a group of pairs within that cable. For example, 06, 151-300 defines the 151st through 300th pair in the administrative cable called "06." See also cable-count.

cache

extra RAM in computer or in server used to hold data requested by workstations; helps data requests execute more quickly.

cache memory

a small high-speed memory placed between slower main memory and the processor. It is a temporary storage area for frequently accessed electronic materials.

CAD

see Computer Aided/Assisted Drafting/Design.

CAD/CAM

see Computer Aided/Assisted Drafting/Design.

cadastral feature

feature on a map created from information from a cadastre; a boundary showing property ownership is a cadastral feature.

cadastral maps

these maps show the spatial relationships among the parcels of land on the cadastre. Typically, these maps record property boundaries and do not show physical features such as lakes or rivers. They are based on a plane survey done at a large scale; the curvature of the earth does not enter into the mapping process. Plate V is a cadastral map.

cadastre

a record of interests in land; an official list of property owners and their landholdings. Land ownership rights are often used as a basis on which to assess taxes.

CADD

see Computer Aided/Assisted Drafting/Design.

calculate

in a GIS, there are typically a number of different styles of calculations that may be performed; these might include arithmetic and Boolean calculations as well as capability to calculate the bearing with respect to true north between any two locations. Others might be the calculation of the average slope of a region or various calculations associated with digital terrain models. Some of the arithmetic calculations may be quite complicated, as in transforming a base map from the grid of one map projection to that of another map projection.

calibration

this common sense word is often used in a mapping context to refer to the idea of aligning a raster or vector data structure with a geographic coordinate system such as latitude and longitude. It is also used in aerial photographic calibration of camera to align lenses and minimize distortion resulting from vibration in flight and other causes. In Plate VII bus routes are aligned with geographic coordinates.

call

a request for a connection to a network.

call accounting

the processing of data on individual calls in a packet-switched network.

call request packet

packet of information containing call user data and information on network facilities sent by the originating Data Terminal Equipment in a packet-switched network.

call user data

user information in packet-switched networks.

called channel

a channel that can receive but not originate calls in a Local Area Network.

calling channel

a channel that can originate calls but cannot receive them, in a Local Area Network.

CAM [1]

Computer-Assisted Manufacturing; Computer-Assisted Mapping; see Computer Aided Drafting/Design.

CAM [2]

see Common Access Method.

CAM [3]

see Content-Addressable Memory.

CAMA

see Computer Assisted Mass Appraisal.

Cambridge Ring

a Local Area Network originated in England; employs an empty slot ring packet circulation pattern.

camera

recording device used to capture visual images on photographic medium (as distinct from a sensor). Aerial photographs, obtained using cameras mounted on satellites or on airplanes, are indispensable to planners, environmentalists, federal and state agencies, and cartographers for obtaining inventories of the earth and its resources. Multilens cameras can record simultaneous views of the same parcel of land; panoramic cameras produce high-resolution photos of a broad expanse of land. Plate III shows an example of an oblique camera shot of San Francisco Bay.

camp-on

any Local Area Network that permits users to wait in line for busy resources and then connects them on a first-come, first-served basis when the resource is free to use.

capacitor

a device used to electrically strengthen cable transmission.

capture

process of freezing and digitizing a video signal; also, frame-grabbing.

card image

stored binary representation of the hole pattern on a Hollerith card.

card module

printed circuit board.

cardinal direction

as on a compass card, the four principal directions: north, east, south, and west.

Carriage Return, CR

the "return" or "enter" key on the QWERTY keyboard. Hitting the carriage return positions the cursor at the left margin of the display terminal.

carrier [1]

electronic equipment that allows more than one customer to be served from the same cable pairs.

carrier [2]

continuous transmission signal which carries a second, information-carrying signal.

carrier detect

a control signal (RS-232) on pin 8. It shows whether or not a local modem is receiving a signal transmitted from a remote modem.

Carrier Sense Multiple Access, CSMA

a contended access communications protocol for sending packets in a Local Area Network. Some variants of CSMA may afford collision avoidance (CSMA/CA) or detection (CSMA/CD), as well.

Cartesian coordinate system/graph

customary x-y axis used to uniquely coordinatize all points in the Euclidean   plane.   A   scheme   devised   by   French philosopher/mathematician, Réné Descartes. The simplicity of the (x,y) coordinatization structure enables it to be extended easily into higher dimensions. In higher dimensions (n dimensions), the coordinate "pair" of the plane is often referred to as an "n-tuple."

cartogram

diagram derived from a base map in which areas or distances are distorted to match some data set. A map of the United States with areas distorted according to population size is an example: in it, Nevada would be represented as an area smaller than New Jersey because the population of Nevada is smaller than that of New Jersey (even though the land area of Nevada is larger than that of New Jersey).

cartographic model

a flow diagram illustrating the various decisions and analyses involved in creating a map from base images and data.

cartography

the art and science of the creation, production, and study of maps.

cartouche

ornate, scroll-like outline surrounding a map legend. Not generally regarded as being of value in digital maps; the content of the legend is more important than the outline of the legend box.

cartridge

erasable storage unit.

cartridge font

cartridge that can be inserted into a printer (usually a laser printer) to add to the printer's capability to produce interesting fonts.

cascade

a style of arranging windows on a computer screen in which the windows sit on top of each other in a stack; as contrasted with tiling, in which windows are adjacent to each other and do not overlap.

catalog (catalogue)

collection of information about entire data sets.

categorical data

discontinuous data that can be mathematically interpolated only; the interpolated values have no sensible meaning--one cannot interpolate between apples and oranges in a meaningful way (leaving out interests of those involved in biological engineering and grafting of trees).

category

in feature mapping, one might partition a class into smaller non-overlapping, but exhaustive, subclasses called categories. For example: the set of all streets in Ann Arbor might be subdivided into three categories--those that are north-south in direction; those that are east-west in direction; and, all others.

Cathode Ray Tube, CRT

television-like vacuum tube in the monitor of a computer used for the visual display of data at a terminal or work station; thin streams of high-speed electrons are projected onto a fluorescent screen to produce the luminous spots that form an image.

CATV

see Community Antenna Television.

CCD

see Charge Coupled Device.

CCITT

Comité Consultatif Internationale de Télégraphique et Téléphonique--Consultative Committee on International Telephone and Telegraph - a technical committee of the International Telecommunications Union, a United Nations organization in Geneva. This committee sets international communications recommendations concerning standardization of data interfaces, modems, and data networks.

CCS

see Common Channel Signaling.

CCT

see Computer Compatible Tape.

CCTV

see Closed Circuit Television.

CCU

see Communications Control Unit.

CD

see Carrier Detect.

CDMA

see Code Division Multiple Access.

CD ROM

see Compact Disk, Read-Only Memory.

cell

in a GIS context, a collection of graphic elements grouped and named as a single item; one value in a raster (array) that corresponds to a particular location on the earth or an abstract space.
Standard cell sizes used by Defense Mapping Agency for digital terrain data (DMA-STD format of DTED equivalents): 1=size 1 degree by 1 degree=scale 1:250,000; 2=size 15' by 15'=scale 1:62,500; 3=size 7.5' by 7.5'=scale 1:24,000.

Also, a geographic area within which a cellular mobile telephone system can make contact with a particular base station for transmitting/receiving purposes. When the mobile phone leaves one cell, it is handed off to the base station of an adjacent cell.

cell library

an organized collection of cells.

cell size

the size, on the ground, of the area represented by one cell in a raster; a cell size of 10 meters means that the cell in the raster represents an area on the ground that is 10 by 10 meters square.

cellular telephone system

mobile telephone system providing service through a network of interconnected, low-powered base stations, each of which serves an area called a geographic cell; offers capability for global extension of mobile telephone service.

census tract

partition of the U.S.A. into small areas by the U.S. Bureau of the Census. Each area represents a neighborhood of similar economic, sociological, and demographic characteristics; a typical tract might have a population of about 4000.

centi

metric prefix for one-hundredth, 10 to the -2 power.

centile

units which partition a distribution into 100 equal segments--as in percentiles.

central meridian

north-south meridian of a map projection on which the map is centered.

central office

a building where switching of telephone calls is accomplished and where all main feeder cables originate. Also: central exchange.

central office area

a geographic area served by a central office or wire center. A central office area is synonymous with a wire center area and is the center where all calls are switched to other exchanges or to toll facilities.

Central Processing Unit, CPU

part of a computer holding the circuitry to control and manipulate instructions to the computer. The major controlling unit of a digital computer; the CPU decodes and executes instructions, performs certain arithmetic operations, and controls use of memory and other internal functions; also known as the computer's "engine."

centralized processing

execution of all of a system's automated processing capability from a central location.

Centrex

a service for customers with many stations that permits station-to-station dialing, one listed directory number for the customer, direct inward dialing to a particular station, and station identification on outgoing calls. The two forms of Centrex are Centrex-CO and Centrex-CU. In Centrex-CO, switching functions are performed at the central office. In Centrex-CU (Customer Unit), switching is done at the customer's premises.

centroid

a point within a closed figure that may be viewed as its "balance" point; the sum of the displacements of points in the figure from the centroid is zero. Centroids give one measure of centrality for a region. The calculation of centroids is straightforward in one and two dimensions; a number of theorems in calculus deal with centroids in various ways (for example, Theorems of Pappus).

Centronics

manufacturer that standardized interconnections for parallel printers using a 36-pin, one byte-wide, connector.

cerography

wax engraving, common in the 19th century. A process that made maps easier to letter than by hand-lettering. Replaced by stick-up lettering and then by electronic and digital lettering.

CGA

see Color Graphics Adapter.

CGM

see Computer Graphics Metafile.

chain

part of a printer--the printing heads that strike a ribbon to leave an inked impression on paper. Also, in a GIS environment, sequence of coordinates used to define part of a line.

chain coding

way to reduce requirements for storage in a raster.

change image
a way of showing spatial change over time using digital maps--an 1992 raster might be subtracted from a 1993 raster to show which streets have been newly paved in the year 1993.

Change In Plans, CIP
a form used to authorize minor changes from an original job print.

changing scale
basically, a simple manipulation of finding a multiplier to alter a representative fraction of a map as desired; the problem comes in making the corresponding spatial content of the map follow. For this, GIS is very useful; thus, the GIS may invoke generalization and thinning in altering the scale of the spatial content.

channel [1]
digital equivalent of a map overlay.

channel [2]
the smallest subdivision of a circuit (or of a trunk route) by means of which a single type of communication service is provided. One-way transmission occurs within a channel. (CCIT)

channel [3]
data linkage joining a Central Processing Unit to peripheral equipment.

channel [4]
electrical transmission along a "channel" linking two or more points.

channel [5]
channel of a Global Positioning Satellite receiver.

Channel Service Unit, CSU
a unit located on the customer premises that provides a Dataphone Digital Service channel for use with the customer's logic and timing recovery circuitry.

character
a symbol which can be represented in a byte, such as in ASCII code.

character string
a contiguous set of characters or by analogy a contiguous set of bytes (representing those characters).

character-oriented
communications protocols with control information coded in character-length (as opposed to bit-length) fields.

Characters Per Second, CPS
measure of the speed at which a printer (for example) can process characters.

Charge Coupled Device, CCD
a CCD camera is a solid state camera that utilizes a CCD to digitize acquired images; CCD cameras read pixel brightness serially on a line by line scan pattern using optical fibers as sensors.

Charge Injection Device, CID
camera system device using photosensitive capacitor elements.

charts
maps designed for navigation are called charts; often the actual course of the ship or aircraft is marked right on the chart. Nautical charts have been in use for centuries. Aeronautical charts are more recent; some are designed for visual flying and others are designed for instrument-guided flight.

Chemical Vapor Deposition process, CVD
a method of making optical fibers from silica and other glass-forming oxides and dopants.

chiaroscuro
graphic method to realistically model shapes by substantial variation in gradation of light and dark.

chip
a small piece of material on which a complete semiconductor device has been built; the device can perform a single simple function like a transistor used as an amplifier, or be a complex integrated circuit replacing thousands of discrete components. Presently, most DOS microcomputers use chips developed by Intel Corporation; other chip manufacturers include Motorola, MIPS, Intergraph, AMD, Apple.

chokes and spreads
means of producing special effects using photographic processes; many of these special effects, such as shadowing are built-in, easy-to-use, features of GISs.

chord
in the plane, when a line intersects a circle, that part of the line that is interior to the circle (on the same side of the circle as the circle center).

chorograph(ic) map
map representing a large geographical region such as an ocean or a continent.

choropleth(ic) map
portrayal of a statistical surface by areal symbols in which the goal is to symbolize the magnitudes of the statistics as they occur inside boundaries of areal units such as census tracts. It is a map in which adjacent areas may not be close in value. The lower right-half of Plate II is a choropleth map.

chroma (Munsell)
the third designation in Munsell specification of color. Neutral gray is zero.

chromatic aberration
image imperfection in an optical system, caused by light of different wavelengths following different paths due to dispersion.

chromatic coordinates
coordinates of x, y, and z represent fractional amounts of primary colors in colors of the Commission International de l'Eclairage system.

chromatic dispersion
dispersion or distortion of a pulse in an optical system due to differences in wave velocity.

chromaticity diagram
a graph of chromatic coordinates; Maxwell diagram.

CID
see Charge Injection Device.

CIE
see Commission International de l'Eclairage

CIP
see Change in Plans.

CIR image
see Color InfraRed photography.

circle of illumination
when the Earth is viewed as a sphere, the great circle that abstractly partitions the sphere into the light half and the dark half; on the equinoxes, only, this circle passes through both north and south poles; on the summer equinox in June, this circle is as far as it will get from both poles, with the north pole in the light half and the south pole in the dark half. On the winter solstice in December, this circle is as far as it will get from both poles, with the north pole in the dark half and the south pole in the light half.

circuit
the complete path, composed of transmitting and receiving channels, between two end terminals that enables one-way or two-way communication to take place; it may be formed from a physical wire (cable), microwave path, satellite transmission, or fiber optic cable.

CISC
see Complex Instruction Set Computer (Computing).

CL
see Cutting Length.

cladding
layer of glass in a fiber optic cable surrounding the fiber core of the cable and within the protective outer covering of the cable.

clarity

on a map, transmission of information is clear when the elements that make up the map are clear and legible and when they are arranged in a balanced manner to aid communication. Thus, legend boxes that are frilly impede communication because they draw the eye to focus on an element of the map with no content. Maps that contain too many symbols produce a cluttered appearance; clutter detracts seriously from the transmission of information. A map that is artistically elegant, with clean simple lines and symbols that are appropriately scaled, is usually effective in transmitting information clearly. The Cover Plate on this book is quite clear; Plate II contains much information, but might strike some as a bit cluttered (visually).

Clarke ellipsoid of 1866

see ellipsoid.

class

generally, a collection of objects sharing some characteristics. May be specialized according to underlying data structure (raster or vector) or according to programming (object-oriented) type.

class interval

in partitioning a set of data into classes, the spacing between elements of the partition may be chosen equally, in increasing or decreasing sequence, or in any of an infinite number of other ways including an irregular pattern that seems to make sense for some reason not related to the abstract notion of forming a partition of the data. The maps below show three different maps shaded using legends with three different class intervals for the same underlying data set. How one chooses to partition underlying data has a great effect on consequent transmission of visual information.

classification

separation of a group of objects into subgroupings, generally that are mutually exclusive yet provide exhaustive coverage of the original group. Taxonomy (science of classification) is an early stage in any scientific investigation. In a GIS context, the word often refers to a grouping of cells of the same type in a raster data structure. The theme on which to group might be spectral characteristics of the cells or any of a host of other themes. In a mapping context, it is useful to partition the infinite number of projections of the earth-sphere into the plane; ways to do this are a source of considerable research interest. There are some useful categories of maps that seem to work well in various contexts: azimuthal, conical, and cylindrical; or, conformal and equal area; as well as a host of others.

Clear-To-Send, CTS

a modem interface control signal indicating readiness for telecommunications to begin.

client-server computing

a system in which much data and programming code resides on a networked host processor (called the"server") that handles the bulk of data processing; a desktop computer (called the "client") provides the user interface.

climate

object of the science of accumulated atmospheric conditions; as distinct from weather (which is local in time and sometimes in space), climate studies overall, broad trends gleaned from systematic record-keeping and analysis of the continuing record.

climax vegetation

the eventual stable grouping of tree types that will dominate in at a given latitude and soil type, after a succession of various less dominant forms of vegetation.

clinographic curve

curve formed in two dimensions by scaling various values.

clip

extraction of data, that reside within the the boundary of features in coverage Y, from coverage X. Clips might be either inclusive or exclusive. Plate II suggests data from one coverage placed adjacent to data from another coverage.

clipboard

temporary storage location used to transfer graphic and textual information between processing applications.

clock

source of timing signals used in various forms of telecommunications.

closed-circuit television

a service often present on broadband networks.

closure [1]

any of several types of housings that can be used to enclose a cable splice.

closure [2]

mathematical systems often exhibit closure; two positive numbers yield a third when added--the set of positive numbers is thus "closed" under addition.   Mathematical closure is a desirable property; in applications it ensures predictability; it is an underlying theoretical requirement for success in digital mapping. A closed user group, CUG, a set of terminal users who do not accept calls from sources outside their group, offer a telecommunications realization of the notion of "closure."

clump

in a vector data structure, a set of contiguous lines, nodes, or polygons merged as a single symbol. In Plate IV students are assigned to the nearest bus stop which is marked with a symbol. The size of the symbol represents number of students at each stop; clumping determines the size of the symbol.

cluster

in reference to computer terminals (and related equipment), a grouping in a single location.

cluster analysis

a statistical method of reducing complexity of data by grouping similar objects in classes and then working with the classes rather than with the individual objects. Often it is useful to work with natural partitions suggested by the data themselves rather than with an artificial classificatory structure.

cluster control unit

equipment that controls the input/output of a cluster of display stations.

cluster labeling

identification of clusters following automated image interpretation.

cluster map

raster image in which categories of variables generally exhibit uncontrolled classification; such maps are not analytical tools for additional analysis because multiple spatially coincident raster data structures have been reduced to a single representative one and numerical comparisons are therefore not accurate.

clustering

tendency of measurements to form clusters in a multidimensional space. Also see agglomeration.

CMIP

see Common Management Information Protocol.

CMIS

see Common Management Information Services.

CMOS

see Complementary Metal-Oxide Semi-conductor.

CMY

see Cyan-Magenta-Yellow.

CO

see Central Office.

Coastal Zone Color Scanner, CZCS
satellite multispectral scanner used to sense ocean water color as a function of plankton and other water-borne materials.

coastline
Boundary between land and water. At a small scale, a coastline is often easy to map. At local (larger) scale, however, difficulty often arises because it can be hard to determine where land ends and water begins. Rivers may feed into swamps and marshlands. All contain water, but one may be navigable and the other not. One might not support trees and the other might. Definitions of coastline become blurred at large scales. Mappers should bear this in mind, particularly when relying on field evidence when making boundary decisions. The coastline in Plate III is quite clear (the water is dark); this oblique photo of the San Francisco Bay area is of relatively small scale.

coaxial cable
network wiring media; varies in thickness depending on transmission requirements; signals transmit on inner copper conductor which is packaged in plastic insulation; the insulation is encased with a foil or copper shield. It is a transmission of wide bandwidth and low susceptibility to interference.

coaxial converter
a protocol converter used between control units.

CODASYL
see COnference on DAta SYstems Languages.

code
a set of logically consistent and unambiguous rules that characterize the way data is represented. Two common codes in the electronic environment are ASCII and EBCDIC.

Code Division Multiple Access, CDMA
digital wireless telecommunications method allowing many users simultaneous access to a single radio frequency channel. This is accomplished by allocating unique code sequences to each user across each channel.

CODEC
see COding and DECoding equipment; coder-decoder.

coder-decoder
a device which translates a coded digital signal to/from an analog signal; the coded side usually uses pulse code modulation but sometimes codecs are used for other coded systems.

coding
data may be entered on a coding sheet and manually digitized or entered from a computer keyboard. Data entry of this sort is a laborious and expensive part to compile digital maps. Electronic digitizers are

more reliable and enter fewer errors. Thus, it is wise to know how the data was coded prior to being mapped; generally, electronic is vastly preferable to manual.

COding and DECoding equipment, CODEC
circuits that digitally code and decode voice signals.

coding theory
digital signals, as those sent from satellites for example, are subject to transmission error of various sorts. The mathematics of coding theory can be used to build codes, digital (binary) signals, that include means, in the length of and pattern within a string of digits, to detect errors and to correct them. Codes of the former sort are called error detecting; those of the latter sort, error correcting. Often these self-detecting and correcting codes are based on finding a shortest distance, called the Hamming distance, from the wrong digits to the "right" digits within the code. See Hamming distance.

Generally, coding theory is the mathematical theory of encoding messages so that it is possible to decode them. The more difficult it is for an outsider to decode a message, the better the code. Codes range from the simple codes generated by the cereal box "decoder ring" to exponential codes based on products of large primes. See public-key cryptography.

coefficient
any member of a product; generally, a constant. Thus, the algebraic expression $2x + 3$ is composed of two terms, 2x and 3, and in the first term, 2 is the coefficient of x; 2 multiplies x.   In applications, coefficients often measure some attribute; distance, direction, or other.

COGO
see COordinate GeOmetry.

collateral data
information to aid interpretation of data sensed remotely.

collision
what occurs when two users on the same Local Area Network post data during multiple simultaneous access to data sets.

color
there are various methods for the classification of color.   Two of them are just briefly mentioned below.

Munsell system: a color is identified in terms of hue, value, and chroma, with each of these dimensions partitioned into equally spaced steps along a scale. Taking the various combinations, there are over 1500 colors in this classification.

CIE system: allows the precise specification of any color in numerical terms, based on physical properties of light, such as dominant wavelength (hue), purity (degree of color saturation), and luminosity (lightness or darkness).

Many digital mapping packages offer "infinite" color selection for (ephemeral) maps displayed on the screen. Colors may be created in an additive or subtractive process in which radiation is added or subtracted to create colors by mixing (removal--selective absorption) or scattering. Plate XI employs discrete use of color to show different speeds.

color balancing

adjustment of red, green, and blue to create a particular effect in a printed image from an electronic file.

color composite

a color negative of a satellite image, created by assigning primary colors of varying intensity to reflectance that varies band by band.

color depth

number of data bits represented by each pixel. In an 8-bit processor, each display pixel can be any of 2 to the power 8 or 256 colors; with a 24 bit processor (or three coregistered 8 bit processors) there are 2 to the 24th power or 256 cubed or 16, 777,216 possible colors for each pixel. Also, pixel depth.

Color Graphics Adaptor, CGA

a type of microcomputer graphics display and/or monitor with horizontal scanning frequency ranging of 15.75 KHz. This adaptor has low resolution and a restricted color selection, as compared to others currently available.

Color InfraRed film

see, Color InfraRed photography.

Color InfraRed photography, CIR

photographic images obtained using special film with sensitivity from green through infrared. In an infrared image, normal red becomes green, normal green becomes blue, and normal blue is filtered out. These images are useful in observing change in that which is normal green, such as vegetation. See Plate III.

color scheme

predefined color combinations a GIS uses for its output display and screen elements.

color separation

a layer representing one of the colors to be used in the eventual printing of a color map (or other document).

color table

set of values used by a computer to assign display colors to digital values in stored data.

column

a vertical list of entries in a raster or a matrix. A matrix column represents a mathematical vector. Within a database or a database management system, a series of common element types across records.

COM

see Computer Output on Microfilm.

command

an instruction to a computer system to perform a specified task.

command interpreter

a shell is an example of a command interpreter that executes user input.

command language

set of instructions to control computer actions.

command line

commands user types in at the system prompt to run an application.

command procedure

a file containing commands and data that the command interpreter reads and executes. The command interpreter can accept the file in lieu of the user's typing the commands individually on a terminal.

commensurable

divisible by a common measure; problems associated with incommensurability often arise in measurements along diagonals in rectangular coordinate systems.

Commission International de l'Eclairage, CIE

This commission has devised a procedure for numerically representing the characteristics of a color based on dominant wavelength, luminosity, and purity.

common access method, CAM

a set of compatibility standards for SCSI peripherals.

common carrier

organization that provides communications services to the general public. It might be a private utility company or a governmental agency; generally, either is subject to local, state, or federal regulations concerning data communications.

Common Channel Signaling, CCS

CCS7 (based on CCITT No. 7 standards) is being used in the implementation of ISDN in the U.S.

common profile

(N-S line) a series of digital points (connected by computer applications software) representing a scaled continuous vertical line to depict the geographic slice of terrain elevation data between adjacent cells for a given interval of arc. The edges of the block in Plate IX suggest a profile.

common raster

(W-E line) a series of digital points (connected by computer applications software) representing a scaled continuous horizontal line to depict the geographic slice of terrain elevation data between adjacent cells for a given interval of arc.

Communications Control Unit, CCU

a communications computer associated with a host mainframe that handles a variety of communications functions.

communications networks

these networks move data from one physical location to another; interfaces to common telecommunications carriers, both switched and leased-line facilities, are usually part of these networks. Often, all user applications programs and databases are concentrated on one or two large host systems in the network.

communications protocol

means used to control the exchange of data communications across a data communications network.

Community Antenna Television, CATV

a common facility on broadband networks; Local Area Network terminology. Allocation of channels is controlled by network standards of various sorts.

COMPANDOR

see COMPressor-expANDOR.

compass

an instrument powered by the earth's magnetic field and used for navigation on the surface of the earth. The compass needle aligns itself with the total field of magnetic force which is not generally parallel with the meridian on which the compass is located. Thus, there is angular variation between magnetic north, where the compass needle points, and true north, where meridians intersect. This angular variation is called magnetic declination.

compass rose

pattern on magnetic compass-card showing N-S, E-W, orientation is often transferred to navigation charts--when it is, it's called a compass rose. Some are simple intersecting arrows; others are arrows surrounded by elaborate designs.

compasses

instrument used to draw circles; one end of this hinged, V-shaped instrument is fixed and the free end, equipped with a writing tip, is free to swing in a full circle. An easy but not too accurate way to draw circles and circular arcs. Many GISs have facilities to draw circles more accurately and to position them anywhere on maps.

compensated optical fiber

an optical fiber in which the refractive index profile has been controlled in such a way that the light rays travelling the short route through the core arrive at the destination at the same time as the light rays travelling through longer paths in the outer material and as those skewed rays following a helical path around the cylindrical cable. This greatly reduces modal dispersion.

compilation (map)

the gathering together of information and measurements to compile a new or revised map according to predetermined standards. Sources for such information might typically include maps, charts, field surveys, remote sensing, or census data. Plate II has been compiled from a number of these sources.

compiler

an internal computer translating program that converts a source code language program into object language; many source code programs must be "compiled" into machine code (the binary representation of the program) for execution by a computer while other programs are "interpreted." Also, one who compiles data to be used in a variety of situations, such as in map-making.

complement [1]

a unit of wire pairs in a telephone cable or at the main distributing frame, usually consisting of either 25 or 50 pairs per complement. The pairs and complements of the unit are numbered sequentially.

complement [2]

in set-theoretic terminology, the complement of a set A within a universe U is the set of all elements in U that are not in A.

Complementary Metal-Oxide Semi-conductor, CMOS

an integrated circuit technology.

complex correlation

process permitting comparisons of data from maps representing different time periods.

Complex Instruction Set Computer, CISC

a computer architecture characterized by a large number of complex instructions implemented mostly in microcode; opposite of RISC.

composite cable

more than one type or gauge of wire in a single common cable sheath.

composite color raster

raster in which each cell contains color information representing one of 256 colors (in an 8-bit processor). The quality of such a display is usually quite good, approaching photographic quality.

composite color video
   color video output of a video camera or VCR in which all color information is contained within one color composite signal.

composite loopback
   a diagnostic test.

composite map
   a single map sheet formed by joining a number of separate images or maps. Plate II is a composite of three separate images.

composite signal
   the output signal of a concentrator or multiplexor.

compressed video
   video images processed to remove redundant information.   This strategy reduces the bandwidth required to send such images over a telecommunication channel.

compression
   of data--reduction of the number of bits required to represent data; of bandwidth--analog compression that reduces the width of the band needed to transmit a signal.

compression dehydrator
   device which dries air to specific humidity levels.

compressor
   equipment used to execute analog compression.

COMPressor-expANDOR, COMPANDOR
   a device for improving the signal-to-noise ratio and signal-to-crosstalk ratio of a communication link, and for decreasing the absolute levels of noise and crosstalk when no signal is being transmitted; consists of a "compressor" at the transmitting end of the link and an "expandor" at the receiving end.

compromise equalizer
   an equalizer that can be set to meet a variety of operational constraints.

Computer Aided Drafting/Design/Drawing/Dispatch, CAD
   software that supports various engineering design considerations. It may include various interactive graphics displays and may offer strong analytical capability in calculation-oriented problems but generally has only limited, if any, analytical capability in attribute processing (as opposed to a GIS which generally has strong capability in attribute processing).   Variants are CADD, and CAD/CAM (Computer-Aided Mapping/Manufacturing). Sometimes, "Computer Aided Dispatch" as in police department terminology to describe an emergency response system based on a package of automated tools to aid in performing standard police work.

computer architecture
see architecture.

Computer Compatible Tapes, CCT
standard 7.5 inch wide magnetic tapes holding digital data. They may be in nine-track or seven-track format--the number of parallel data recording tracks.

computer graphics
electronic systems designed to display and to manipulate visual material; they generally do not have analytic capabilities as do GISs.

Computer Graphics Metafile, CGM
a graphic image exchange standard; ANSI: x3.122-1986; ISO: 8632-1986.

computer network
two or more Central Processing Units and peripherals linked by telecommunications.

Computer Output on Microfilm, COM
system that stores images on computer-generated microfilm; images can be printed on a microfilm printer.

computer-assisted cartography
use of the computer, in an interactive manner, in the creation and production of maps.

Computer-Assisted Mass Appraisal, CAMA
automated analysis of property values and related items as a basis for tax assessment.

computerized tomographic scanners
equipment used in the medical sciences to create imagery of humans similar to the imagery satellites and GISs create of the Earth; CAT-scans.

COMSEC
specifications for data encryption and decryption for SECure COMmunications.

concatenate
to link records or files in a series for sequential processing or storage.

concentration
hierarchical collection of data from several lower speed transmission lines to be fed into a single higher speed transmission line.

concentrator
equipment used to partition a data channel into two or more channels of lower speed. Although the average speed is lower in each of the many channels, data throughput is maximized by allocating space in the channels dynamically according to demand for channel use.

concurrency

when more than one process is working on a problem; can occur with either multiprocessing or parallel processing.

conditional map element

elements of the map subject to societal constraints, such as rights of access, property lines, and so forth. See Plate V.

conditioning

the addition of equipment to a data transmission line designed to improve transmission characteristics.

conductor

the individual wires in a cable (copper, aluminum, or other).

conduit

a pipe, tube, or compartmentalized structure placed underground to form ducts through which cables can be passed. Conduit is also used inside building walls or floors for the same purpose.

CONECS; CONNECS

see CONnectorized Exchange Cable Splicing.

COnference on DAta SYstems Languages, CODASYL

the committee that designed COBOL .

configuration

the make-up, pattern, contour, shape, or composition of a particular map feature. The configuration of the buffer zones in Plate X serves to suggest which parts of South Carolina lie within ten miles of an Interstate Highway. The arrangement pattern of computer hardware and peripheral devices forming a computer system.

conflation

set of techniques to align arcs of one coverage with those of another.

confluence

the flowing, meeting, joining or coming together into one drainage pattern of streams, rivers, glaciers, oceans, et cetera. Plate IX shows clearly two streams coming together uphill from the mouth of the stream.

conformal projection

a conformal projection is one in which angle is preserved; angular relations are correct relative to the compass rose. Preservation of angle is at points, so that correct shape is preserved but only locally. Broad regions show substantial distortion. A conformal map is useful for analyzing motion involving angular relationships, such as navigation and meteorology. Often topographic maps are made on a conformal base map. Some common conformal projections are the Mercator, the transverse Mercator, Lambert's conformal conic with two standard parallels, and the stereographic.

Examples of the Mercator (see entry on cylindrical maps) and the Lambert (Figure 13 and Figure 14) are given in this book. The Lambert projections below illustrate clearly that shape is good locally but not globally. Lambert's conformal conic has two standard parallels and equally spaced, straight meridians that meet the parallels at right angles. On the map of the co-terminous United States in Figure 13, the standard parallels are close to 45 degrees and 30 degrees north latitude. Near those lines area distortion is slight. Indeed, when the whole world is viewed from this perspective with these two standard parallels (Figure 14), area distortion is only slight for the landmasses near these parallels. This projection offers good directional and shape relationships for an east-west zone. Area distortion away from the standard parallels is severe--consider South America, which appears greatly elongated, and Antarctica which is massive (Figure 14). Thus, where one chooses to truncate, or crop, a map projection determines how realistic the final map will be (when compared to the globe).

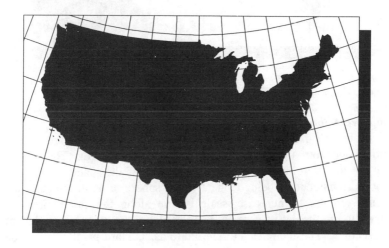

Figure 13. Lambert conformal conic. Standard parallels positioned to make the co-terminous U.S.A. appear true.

In Mercator's projection (see entry), there is considerable distortion as one moves away from the equator. The parallels and meridians are perpendicular in the Mercator; all lines of constant compass bearing (rhumbs) appear as straight lines and this is a great advantage for navigators. This projection is good for little else because distortion away from the equator is severe.

The transverse Mercator has a zone on either side of a central meridian (rather than the equator, as in the Mercator) around which shape is good and away from which distortion is severe. It too is conformal and is the base for the Universal Transverse Mercator (UTM) plane coordinate system.

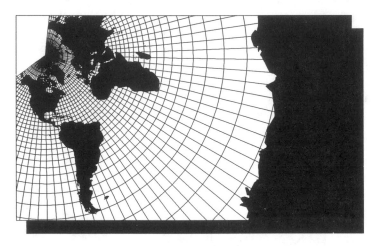

Figure 14. Lambert conformal conic. Standard parallels positioned to make the co-terminous U.S.A. appear true--much of cone visible.

conic projection

conceptually, a conic projection is the projection of the earth sphere onto a cone from a point (often) at the center of the earth. The cone is then slit and unrolled--developed--into a part of the plane. The cone may be envisioned as being tangent to the sphere or as intersecting the sphere. In the former case, scale is true along the small circle of tangency; in the latter case, it is true along the parallel circles of intersection. These circles are referred to as standard parallels and true conic projections should state whether they are based on one or two standard parallels (see Albers projection and conformal projection for examples); when more than one cone is used to develop a projection, it is referred to as a polyconic projection.

Figures 15 and 16 show an equidistant conic: one on which the graticule is laid out mathematically so that all parallels are evenly spaced on a flattened cone. This strategy ensures that distances measured from the center are true. The appearance of the map of the co-terminous U.S.A., and at other parallels close to the center, is reasonable (Figure 15); at the global scale there is considerable distortion (Figure 16) away from the center.

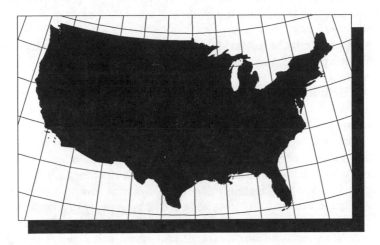

Figure 15. Equidistant conic--positioned to make the co-terminous U.S.A. appear true.

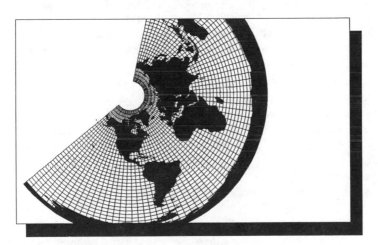

Figure 16. Equidistant conic--positioned to make the co-terminous U.S.A., and other landmasses at that latitude, appear true (globe-like). Conical structure is apparent.

conjugate principal point
along a flight line, position on a photograph of the principal point of the adjacent photograph.
connect time
the length of interval (time) during which the user was on-line (connected to the computer).

connect-through pairs

pairs idled when service is discontinued but which remain connected to the customer's premises. The pairs are usually, but not invariably, listed as assigned if expected to be idle less than 60-90 days.

connecting blocks

blocks are designed to provide cross-connections between feeder and distribution plant.

connection

an established data communications route through a communications network.

connectivity

a mathematical concept that offers ways to measure the extent to which various mathematical objects are linked to each other. Graph theory is a branch of mathematics which studies connections among objects. It has applications in virtually all fields of human study, and because graph theoretic problems have arisen in so many diverse fields, there are many names for the objects and connections. The most common are points and lines, vertices and edges, and nodes and arcs. This last is most common when the direction of connection is important (directed graph or digraph). Some engineers still refer to graph theory as network theory. Also see graph theory, topology.

In a GIS environment, connectivity refers to a capability to identify points or areas that are joined to other points or areas by lines. In graphics, it refers to mere physical adjacency. In a data base, involving the management of public utilities (for example), items might be linked through source Item of Plant Identification numbers.

CONnectorized Exchange Cable Splicing, CONECS

cable where connectors are pre-installed on one or both ends of the cable at the factory. Connectors are simply pressed together with the use of a special hand tool to complete the splice.

connectors

various types of electrical devices used to join wire.

console

equipment used by computer personnel to monitor various aspects of computer performance and to allow the operator to communicate with the computer.

constant

a fixed value, in contrast to a variable value.

constraints

limits within which a problem is to be solved; often expressed as equations or inequalities to which other equations describing real-world processes are subjected. For example, consider all locations in the northern hemisphere of population greater than 1 million--the constraint here is "the northern hemisphere" which can be captured as "set of points with latitude between 0 and 90 degrees north."

contact print

a photographic print made directly from a negative or positive in contact with sensitized paper or film.

contactor

attachment device that joins cable pairs to a pressure pipe system.

contended access

in Local Area Networks, as opposed to explicit access. Stations share access to use the network on a first-come, first-served basis.

content-addressable memory, CAM

also known as associative memory, a memory with comparators on each cell to allow a parallel comparison of the input data with all the memory contents.

contention

telecommunications facility allowing n (n>1) terminals to compete for m computer ports (m<n) on a first-come, first-served basis.

contiguity analysis

analysis concerning relationships between and among adjacent polygons. The buffers in Plate X suggest a transport relationship between South Carolina counties.

contiguous

physically adjacent and/or consecutively numbered units of data. The drainage basin in Plate VI is formed from data in 36 contiguous tiles.

Continuing Property Record, CPR

an accounting inventory and record of telephone company property. A mechanized record of property items (plant). The CPR system identifies and summarizes plant quantities by:
1. CPR group number; 2. Field Reporting Code/Plant Account Code;
3. Division/district; 4. Percent ownership; 5. Year placed (mortality);
6. Ad valorem tax code.
Digital maps, such as the one in Plate V, show a visual display of this kind of record.

continuous data

values in a raster data structure are continuous if interpolation of values always leads to results that have meaning.

continuous presence video

video teleconference offering simultaneous and continuous pictures of all participants.

continuous tone

smooth transition in tone as on a photograph.

continuously variable

in describing analog transmissions, a variable capable of assuming any one of an infinite number of values, differing from each other by an arbitrarily small quantity.

contour

an imaginary line, all points of which are at the same elevation above or below a specified reference or datum surface (usually mean sea level).   As a verb, the process of locating contour lines in a data set. In terms of linkages to mathematics, a contour line is a level line of a surface which may be found using methods of the calculus.   There are various types of contours: an index contour has numerals adjacent to it showing elevation; an intermediate contour shows the spacing interval between successive index contours; and, supplemental contours give some sort of extra information. Plate I right side, and Plate II left side, have contour lines, suggested or present, to depict different elevation levels of abstract (Plate I) and natural (Plate II) surfaces.

contour interval

the vertical distance between the elevations represented by adjacent contour lines on a contour map.

contour line

see contour.

contour map

a map showing the configuration of a surface by means of contour lines drawn at regular intervals of elevation; a topographic map. See contour.

contrast

difference between bright and dark (intensity) elements of a map, photograph, or other display. See Plate IV, for example.

contrast enhancement

see contrast stretching.

contrast stretching

contrast stretching involves extending the given range of recorded digital values to the full range available. This procedure increases the amount of contrast on the image. Also called contrast enhancement.

control character

a printer or carriage character that does not print out; used to start, stop, or modify a function.

control point

in mapping, a point with given horizontal (plane) position, (x, y) and a known surface elevation z measured above some base. This point is used in estimating unknown elevations of other locations within the area to be mapped.   Control points are used as fixed references to position and correlate map features, and to rubber sheet vector and/or raster overlays.

control point list

list containing paired information of map coordinates and cell coordinates in a raster.

control signal

a modem interface signal to start, stop, or modify a function.

control station

a place from which a traffic control system or remote control signal appliances are operated or from which instructions are issued (RR).

control unit

on an IBM host, the equipment that coordinates the operation of an input/output device and the Central Processing Unit.

controlled airspace

airspace within which some or all aircraft may be subject to air traffic control.

conventional memory

in a DOS environment, the first 640 kilobytes of memory for running applications.

conversion

the changing of one system of measurement to another; the process of changing the form in which digital data is expressed.

cookie cutter

see clip.

coordinate filtering

process used to generalize maps by removing superfluous coordinates.

COordinate GeOmetry, COGO

map data system designed for entry and transformation of land survey data such as bearings, distances, and angles into a coordinate system. See also Cartesian coordinate system and coordinate system.

coordinate pair

any ordered pair of numbers used to position a point relative to coordinate axes in the plane. In mapping, Cartesian coordinates often describe the location of a point, line, or area/polygon in relation to the coordinate system of the common underlying data base.

coordinate system

a frame of reference that uses linear or angular quantities (typically) to designate the position of points with respect to the frame of reference. Examples include rectangular coordinates, oblique coordinates, polar coordinates, or spherical coordinates. See also Cartesian coordinate system.

coordinate transformation
   mathematical mapping of one coordinate system to another.
coordinates
   in mapping, a set of ordered numbers that determines position--
either absolute or relative.   The tiles in Plate VI have corners with
evenly-spaced coordinates.
copy
   printing term to refer to any graphic or textual material to be
reproduced.
copyright
   legal term referring to protection of written documents.   In the
U.S.A., the copyright office is in the Library of Congress.
Cordless Telephone, second generation, CT-2
   mobile phones connected by Time Division Multiple Access links to
microcell base stations. A wireless payphone on which it is not possible
to receive calls.
Cordless Telephone, third generation, CT-3
   similar to CT-2, except that it is possible to receive calls.
core
   a tiny ring of magnetic material strung on a fine wire matrix,
forming the random access memory of early computer designs.
core memory
   the main memory of a computer; this is the most accessible
information storage unit of the computer. It provides high-speed access
to its content during program execution; RAM refers to core memory in
a microcomputer environment.
coregistration
   alignment of raster or vector data structures such that the geometries
of internal features coincide.
Corporation for Open Systems, COS
   consortium of computer vendors and end users whose charter is to
promote the implementation and acceptance of OSI communication
protocols and to assure compatibility through conformance testing.
correlation
   refers loosely to the interdependence between two sets of numbers.
When this interdependence is characterized as a mathematical relation,
it is such that when one set changes, the other does, too.   When an
increase in one set (or decrease) causes an increase (decrease) in the
other, the correlation is positive; when an increase (decrease) results in
a decrease (increase), the correlation is negative.
corridors
   see buffers.

COS

see Corporation for Open Systems.

co-terminous United States

the 48 states; Alaska and Hawaii are not co-terminous.

coverage

basic unit of data, a single map sheet--ARC/INFO; same as theme, channel, layer, level in various mapping programs. In Plates VII and VIII the layer of school information is placed on the layer of the street network.

CPR

see Continuing Property Record.

CPS

see Characters Per Second.

CPS code

in a global positioning system environment, so-called civilian code.

CPU

see Central Processing Unit.

CR

see Carriage Return.

crab

the condition caused by failure to orient a camera with respect to the track of the airplane; in aerial photography, crab is indicated by the edges of the photographs not being parallel to the air-base lines (yaw).

CRC

see Cyclic Redundancy Checking.

critical angle

the angle with the normal at which total reflection of a ray incident to a surface first occurs.   With respect to total internal reflection in an optical fiber, the critical angle is the smallest angle made by a meridional ray that can be totally reflected from the innermost interface of the fiber. It therefore determines the maximum acceptance angle at which a meridional ray can be accepted for transmission along a fiber.

cross box (cross connect box)

an item of outside plant hardware in which jumper wires are used to connect feeder pairs to distribution pairs. In is also called an X-Conn.

cross connect box

see cross box.

cross connection

a means of semi-permanently connecting cable pairs from one group to cable pairs in another group. Cross connections can be made on terminating frames or in enclosures designed for that purpose which are called "cross-connect facilities" or "cross boxes."

cross hatching
    an areal pattern often consisting of two intersecting sets of parallel lines, used for shading map regions.
cross section
    in mapping, a section perpendicular to a centerline.   When a sequence of sections is taken at regular intervals or break points, they can be used to determine new or altered contour plots or volumes. The edges of the block in Plate IX show a cross section of terrain.
cross-connect terminal
    type of terminal placed between underground cable coming from the CO (called feeder cable) and distribution cables. Also called B-Box or 80 terminal.
crosstalk
    undesired transfer of a signal from one circuit to another.   The former is referred to as the disturbing circuit; the latter as the disturbed circuit.
CRT
    see Cathode Ray Tube.
CSB
    (obsolete) an old type of distribution terminal which is both Ready Access and Fixed Count.
CSMA
    see Carrier Sense Multiple Access.
CSMA/CA
    see Carrier Sense Multiple Access.
CSMA/CD
    see Carrier Sense Multiple Access.
CSU
    see Channel Service Unit.
CT-2
    see Cordless Telephone, second generation.
CT-3
    see Cordless Telephone, third generation.
CTS
    see Clear-To-Send.
cubic interpolation
    use of a cubic curve (polynomial of degree 3) to interpolate a bounded, finite, set of points. Offers greater accuracy of fit than do linear fits.   Cubic spline analysis is one often-used technique for interpolation; it is derivative of the concepts behind the cartographer's spline, a slender flexible wooden strip which can be bent and held in place (traditionally using lead "whale" weights) to fit any curve.

CUG

Closed User Group; see Closure.

cultural features

map features constructed by humans. They might lie above, on, or below the earth's surface. Examples include buildings, roads, canals, tunnels, or sewer systems. The roads in Plate XI are cultural features.

cumulative frequency graph

frequency graph based on accumulating areas of regions on the x-axis and plotting ranked values of some geographical distribution on the y-axis; sometimes called a hypsometric curve--area/altitude analysis is an example.

cursor

a movable underlining or highlighting device that indicates a location visible to the machine operator; its coordinates are known by the machine. The cursor in a digitizer is a hand-held device with a cross-hair locator and multiple buttons (a large version of a computer mouse--jokingly, a "rat") that records the position of objects on a map and transfers these locations to a computer file.

cutover or cut-over

the transfer of service from one facility to another. Usually refers to a transfer of an outside service wire, or drop wire, from one cable to another cable pair.

cutting length, C or CL

length of cable placed before splicing.

CVD

see Chemical Vapor Deposition process.

Cyan-Magenta-Yellow, CMY

the subtractive (absorption) color system, of dots of cyan, magenta, and yellow, used by color printers to create full-color images; occasionally termed YMCK or CMYK (to include blacK).

cyclic redundancy check

an error checking routine that uses divisibility characteristics of the transmitted data to detect telecommunications errors.

cylindrical projection

conceptually, a projection of the earth-sphere from its center onto a tangent or intersecting cylinder. The cylinder is slit and unrolled--developed--to produce a map in the plane. In the former case of the tangent circle, a great circle is a line along which scale is true; in the latter case of two small circles derived as lines of intersection, these two small circles are the lines along which scale is true.

In Mercator's projection (Figure 17), a cylindrical projection conceptually that must be developed mathematically, there is considerable distortion as one moves away from the equator. The parallels and meridians are perpendicular in the Mercator; all lines of constant compass bearing (rhumbs) appear as straight lines and this is a great advantage for navigators who can approximate a great-circle route by a sequence of straight rhumbs. This projection is good primarily for navigation.

Mercator's projection is centered along the equator with evenly spaced meridians. It is not an equal area projection; landmasses poleward are greatly exaggerated. Thus, it would not serve as a reasonable map on which to compare visually extent of forested lands in northern Canada and Brazil. It is a conformal projection. Concerns of this sort are important to note when making policy decisions concerning allocations of funding for issues involving, for example, broad expanses of forested lands.

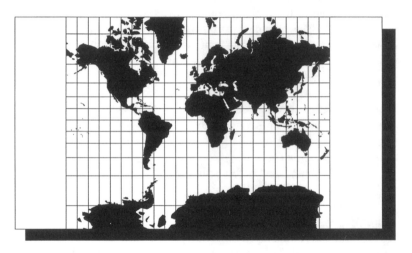

Figure 17. Mercator's conformal cylindrical projection. The Equator is the central line.

Behrmann's cylindrical projection is an equal area projection (Figure 18), so if equal area characteristics rather than navigation and conformality are dominant, one might wish to choose this projection. One "criticism" of the Mercator projection, which is NOT true, is that it "favors" the northern hemisphere. The projection does not; it has the Equator as its central line. One may choose to truncate the projection so that more of the paper map is devoted to the northern rather than to the southern hemisphere. There are good reasons for doing so; most of the world's inhabited lands are in the northern hemisphere. The map in Figure 17 is not truncated in what has come to be a conventional manner; hence Antarctica dominates. This sort of domination is undesirable in many situations. Again, how one chooses to crop a map can influence perception: Mercator's projection is of infinite extent and it is centered on the equator as the great circle to which the developing cylinder is tangent. Those who like it might choose to crop it to suit their interests.

The transverse Mercator is also developed conceptually from a cylinder tangent to a great circle passing through the north and south poles (transverse to the tradition Mercator). It has a zone on either side of a central meridian (rather than the equator, as in the Mercator) around which shape is good and away from which distortion is severe. It too is conformal and is the base for the Universal Transverse Mercator (UTM) plane coordinate system.

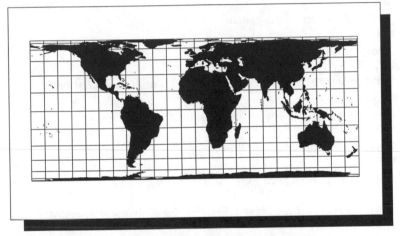

Figure 18. Behrmann's equal area cylindrical projection. The Equator is the central line.

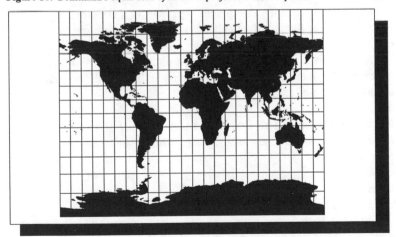

Figure 19. Gall's cylindrical projection.

By far the most familiar projection in this class is the Mercator; while it serves well for navigation, it has unfortunately been used far too often for purposes for which it is not well-suited. There are a number of other cylindrical projections: Gall's projection (Figure 19), land Van der Grinton's, are also useful but have been, no doubt thoughtlessly, misused.

CZCS
see Coastal Zone Color Scanner.

D d

DAA
see Data Access Arrangement.

DAB
see Data Acquisition Board.

daisy chain
the process of interconnecting peripheral units in a serial fashion.

dangle length
minimum length allowed for dangling arcs.

DAP
see Displayable Attribute Processing.

DARPA
see Defense Advanced Research Projects Agency.

data
discrete sets of quantized graphic or alphanumeric information (singular: datum).

Data Access Arrangement, DAA
Data Communications Equipment supplied by a common carrier; all modems built for the public telephone network have integral DAAs.

Data Acquisition Board, DAB
a data acquisition system incorporated on a printed circuit board that is compatible, electrically and mechanically, with a particular microcomputer system.

data bank
the sum total of all related data bases together with a centralized information storage and retrieval system which can be inventoried and accessed.

data base
see database.

Data Base Management System, DBMS
a generic term describing any computer system devoted to the management of large collections of persistent data (see database).

data collection
accumulation of information at one location, in a file or a queue.

data communications
the set of processes, equipment, and facilities used in the transport of signals from one data processing device to another.

Data Communications Equipment, DCE
often used as a synonym for modem. Also referred to as Data Circuit-terminating Equipment.

data conversion
translation of data formats to enable exchange of data from one system to another.

Data Definition Language, DDL
a language used to define the structure of a data base.

data dependency
when data produced by one part of a concurrent job is used by another part.

data dictionary
a database that catalogues map information or the contents of other databases and tables.

data element
see element.

data encoding
conversion of data to machine-readable form.

data entry
process of entering tabular data into a computer.

Data eXchange File, DXF
spatial data exchange format developed by Autodesk, Inc. Files produced in this format are .DXF files.

Data Exchange Format, DEF
spatial data exchange format used by the U. S. Forest Service, on a limited basis, as a means of file transfer with other agencies.

data format
manner in which data are represented and stored in electronic records by field type and field length.

data fragmentation
use of Boolean operations to partition data in relational databases; also, garbage from repeated disk access, causing disk fragmentation.

data granule
part of a data set; a basic useful component, such as a single image.

data integrity
an evaluation of performance in transmitting data. The lower the measure, the fewer undetected errors in the transmission.

data layer
  see layer.
Data Link Control, DLC
  IBM's SNA protocol layer that controls data transmission, error detection, and error recovery between two nodes.
data link layer
  second level in OSI reference model; controls data flow in and out of each network device.   Passes data from the network layer to the physical layer.   Handles transmission and reception of packets, datagrams, local addressing and error detection (only--not error correction).
Data Management Retrieval System, DMRS
  provides the user with the means to access, manipulate, and maintain a data base.
Data Manipulation Language, DML
  defines the operations performed on a data base.
data mask
  data processing barrier that allows only user-selected data values to pass by it.
Data Network Identification Code, DNIC
  a four digit PDN identifying code in a packet switched network. In the United States of America there are a number of such networks; these are listed below together with their accompanying DNIC: Autonet, 3126; Compuserve, 3132; DBS(WUI), 3104; Datapak, 3119; LSDS(RCA), 3113; Marknet, 3136; Telenet, 3110; Tymnet, 3106; UDTS(ITT), 3103; Uninet, 3125; Wutco, 3101.
data PABX
  a PABX used for data whose function is to set up and break electronic connections on demand.
data point
  a record in the computer defined by its (x,y,z) position related to a point of reference = 0.
data rate
  rate at which data is transmitted--expressed in bps (sometime expressed incorrectly in baud).
data redundancy
  protective strategy in which copies of identical data are stored in several different computer locations.
data security
  strategy to prevent unauthorized use of any data, whether or not it has been stored redundantly.

Data Service Unit, DSU

a terminal located on the customer premises for the purpose of accessing the Digital Data System through a standard EIA or CCITT modem interface.

data set

commonly, a collection of data, or just "data."

data signaling rate

see data rate.

data spacing

the distance between centers of digital data elements in any data matrix or evenly spaced array along a given direction (e.g., statistical sample and 200 line screen). The tiles in Plate VI have evenly spaced centroids (as data points).

data stream

set of alphanumeric characters and data bits sent through a channel.

data structure

organization of data and of reference linkages between data in memory.

data system

collection of hardware and software for managing and accessing data.

Data Terminal Equipment, DTE

terminals and computer ports serving as the data source or sink.

data type

in general, the way in which bits are grouped and interpreted; in reference to the processor instructions, the data type of an operand identifies the size of the operand and the significance of the bits in the operand. In a spatial interpretation, often refers to a map layer.

data-over-voice

method of combining data and voice on same twisted pair cables; often on in-house telephones.

database, data base

a structured collection of ordered data, needed by an organization to meet its information processing and retrieval requirements, and critical to a system stored in graphic, textual, or digital form. It is a structured collection of data that has been defined by a particular use, user, system, or program so that it can be expanded, updated, and retrieved quickly; it may be referred to as sequential, network, hierarchical, relational, or semantic--terms which also refer to the types and methods of queries that are supported. It contains at least one file of data and often includes more. Geographic Information Systems interact with data bases so that maps can be dynamically updated over time.

datagram

in packet-switched network, a connectionless single packet message.

Dataphone Digital Service, DDS

a communications service--AT&T leased lines that transmit digital rather than analogue data; thus, modems are not needed.

datum [1]

a level surface to which elevations, or other measurements, are referred.

datum [2]

singular of data.

dB

see deciBel.

dB2

relational database management system of Microsoft Corporation.

dBaseIII, III Plus, IV

PC database management system marketed first by Ashton-Tate and currently by Borland.

dBm

the amount of power relative to that represented by a signal which will dissipate one milliwatt of power into an established resistive load (normally 75, 135, 600 01-900 ohms).

dBmO

the power in dBm measured at, or referred to a point of zero relative transmission level.

DBMS

see Data Base Management System.

DCE

see Data Communications Equipment.

DCS

see Digital Crossconnect System.

DCW

see Digital Chart of the World.

DD

see DeaD pairs.

DDCMP

see Digital Data Communications Message Protocol

DDGT

see Digital Data Group Terminal.

DDL

see Data Definition Language.

DDS

see Dataphone Digital Service; also Digital Data System.

dead ends

incomplete linkages on a digital map.

DeaD pairs, DD

identifies pairs in the cable which are not connected to the Main Distribution frame. There are two types of dead pairs: Idle Dead and Dead Dead. Idle pairs are spliced through but are not energized. Dead pairs are not spliced together.

debugging

detecting and correcting errors in a computer program.

deca/deka

metric prefix for 10 to the power 1.

decentralized processing

processing that is distributed across a network of equipment in different locations; the difference between centralized and decentralized processing is much like the difference between a centralized university library system (with all holdings in one building) and a decentralized university library system, in which one might find a mathematics library and a natural resources library in addition to a main library.

deci

metric prefix for one-tenth, the power 10 to the -1.

deciBel, dB

measure of transmission levels, gains and losses measured on a logarithmic scale such that the number of decibels is equal to 10 times the logarithm to the base 10 of the power ratio.

decimal

in a computing environment, a digital system of ten states, 0 through 9. Mathematically, a decimal system is a closed system in which any element has a unique expression as a polynomial using powers of 10 with coefficients ranging in value from 0 to 9. Thus, 301.27 is a unique representation of the polynomial

$$3*10^2+0*10^1+1*10^0+2*10^{-1}+7*10^{-2}$$

DECnet

trademark for Digital Equipment Corporation's network architecture; a family of software and hardware communications. One feature of a DECnet network is that its nodes can route messages through the most cost-effective path; if a circuit is disabled, nodes route messages through alternative routes.

DECT

see Digital European Cordless Telephone.

dedicated lines

leased telecommunications circuits devoted to a specific application; for example, leased telephone lines could be used for a teleconference network connecting fixed locations.

dedicated system

a facility devoted to a specific purpose.

DEF

see Data Exchange Format.

default

a value of a parameter, device, or code that is assigned automatically if no specific value assignment is made.

deflection angle

measure of the margin of distortion in photogrammetry.

degree

a unit of angular measurement--1/360th of a circle. In a graph, the number of edges incident with a given node; in a graph-theoretic tree, every node has degree two or one (the tips of the branches). In Plate VIII, the tips of the spiders (stars) each have degree one; the central node has relatively high degree.

delimiter

a character that denotes the beginning and end of a character string and, therefore, cannot be a member of the string.

delivery confirmation bit

in an X.25 packet-switched network a bit requesting end-to-end acknowledgment.

DEM

Digital Elevation Model; see Digital Terrain Model.

demodulation

process of retrieval of data from a carrier (opposite of modulation).

demographic map

a map that depicts political or social data, often in combination with each other and frequently based on census data; populations by national political or subnational political units.

demultiplexing

process of splitting a composite signal into its component channels (opposite of multiplexing).

denormalization

process that improves speed of retrieval processing of databases.

densitometer

instrument to measure quantity of transmitted or reflected light.

density slicing

conversion of a continuous graytone image into a sequence of discrete gray tones, one for each of a set of digital ranges.

Department of Defense, DOD

branch of the federal government of the United States of America, executive branch, that handles military affairs. It is responsible for data communications and for selected Local Area Network protocols and standards.

dependent variable

in a mathematical function, the variable that is the output of the function; an element of the range of the function--the y-value in y=f(x) that depends on the input x and the rule that tells how to manipulate x in order to generate y.

depression contour

depicts (concave) areas inside of contour lines that are equal to or lower than the surrounding terrain. In Plate IX, the three-dimensional depressions would be represented by depression contours on a topographic map.

depressionless elevation

portion of a surface with all depressions, such as puddles or holes, filled.

derived map

map formed from combining maps from a master database.

design file

an area set aside on disk or magnetic tape for storage of graphic elements; a specific organizational unit corresponding to a map sheet.

Design Rules Checker, DRC

software that examines a design to make sure it conforms to manufacturing rules.

Deutsche Industrial Norms, DIN

a West German agency that sets engineering and dimensional standards recognized worldwide.

developable surface

a developable surface is one that can be slit and flattened out completely in the plane. This is an important issue in mapping; a sphere is not a developable surface; hence plane maps merely capture some, but necessarily not all, of the characteristics of the globe. One strategy for creating map projections is to choose to project the globe onto a developable surface and then slit the surface and develop it as a region of the plane. Cylinders and cones are developable surfaces that are commonly used for this purpose; an hyperboloid of one sheet is also a developable surface (the shape of nuclear cooling towers) but these do not appear to be used, to date, in map making. A developable surface is one that can be generated by an appropriate motion of a single straight line generator: a cylinder is swept out by a line normal to a plane as the line revolves around a circle in that plane. See various black and white maps in this Part developed from a cone.

deviation

difference between magnetic and gyro compass headings; also a statistical term to describe the difference between, for example, observed and expected values.

device

equipment external to the computer designed for a specific use.

device driver

a program that translates commands from application software, such as a word processor, into instructions that a device, such as a printer, can use to perform a function--such as printing.

DeVice Independent, DVI

not constrained by hardware of a particular environment. TeX is designed to create DVI files from ASCII files of words and mathematical notation that work the same way in various hardware configurations.

device name

identification of a physical device or a logical name that is equated to a physical device name.

device server

hardware used to connect equipment such as dumb terminals and printers to a LAN (independent of the topology of the LAN); it is dedicated to the supervision of commands between attached devices and the network.

DFAD

see Digital Feature Analysis Data.

DFB

see Distributed Feedback Laser.

diagnostic

a program that tests hardware/firmware logic and peripherals and reports any faults it detects.

dial-in line

a connection to a public telephone line.

dialog box

window in software that allows the user to change various settings.

diapositive

positive photographic image on a transparent medium such as glass, or film--mirror image of a photographic negative.

dibit

a set of two bits. Since the digits are binary, there are four possible values: 00, 01, 10, and 11.

dichotomous data

data concerning exactly one subject, which either is or is not present in the larger context--examples: yes/no, 0/1, black/white.

dichroic filter
   an optical filter that transmits all frequencies above, and reflects all frequencies below, a cutoff frequency.

dichroic grating
   in a multispectral scanning sensor system, a grating that separates incoming reflected, from emitted, wavelengths.

differential equations
   a branch of mathematics in which techniques of algebra are used to solve equations, and systems of equations, containing objects (such as derivatives and differentials) from the calculus. Differential equations arise in many calculations involving the earth, particularly when its geographic grid is transformed, and they also are at the base of all logistic growth which is frequently used in analyzing population. They also appear in the rigorous characterization of any phenomenon involving either physical or mathematical waves, resonance, and rhythmic beats.

differential phase shift keying
   a modulation scheme used in selected Bell modems (e.g. 201).

differential positioning
   in the scheme of a Global Positioning System, the relative position of receivers tracking the same signal.

diffraction
   process in which lightwaves are modified by interaction with obstacles; some rays are caused to deviate from their paths by the obstacles while others are not. A redistribution of light rays results from diffraction following the incidence of a lightwave with an obstacle(s).

diffraction grating
   an array of fine, parallel, equally-spaced reflecting or transmitting lines with spacings related to the order of the wavelength of the incident light. Gratings might be ruled grids, spaced spots, or crystalline lattice structures; the resulting patterns of light and dark areas are "diffraction patterns."

digit
   a discrete numerical value; numerical identification of the pulses used to construct a binary code.

digital
   the discrete, numerical, representation of information--contrast with iconic or analog representations.

digital capture device
   optical device to convert hardcopy maps or other analog images into a digital raster form which can then be displayed and analyzed.

Digital Cartographic Interchange Format
spatial data exchange format developed by Battelle Labs for FEMA.
Digital Chart of the World, DCW
a 1:1 million vector map database prepared by the U.S. Defense
Mapping Agency.
digital computer
computer whose operations are base on discrete variables rather than
on continuous ones (analog computer).
digital contour plot
a computer generated plot of digital line data. See Plate IX.
Digital Crossconnect System, DCS
a machine designed to connect T-carrier lines, generally between
central offices.
digital data
records represented digitally in a computer system.
digital data base   •
a set of data maintained in common format(s) that supports one or
more different user systems.
digital data collection
computerized system for storing digital line data in a programmed
format. One example is the Digital Terrain Data Collection System.
Digital Data Communications Message Protocol, DDCMP
a communications protocol used in a DEC environment for
communications between computers.
Digital Data Group Terminal, DDGT
a terminal located in a DDS office that provides the electrical
interface between a DDS channel at the DS-0 level and a full duplex
analog group band transmission system.
Digital Elevation Model, DEM
spatial data exchange format in use by USGS. Also see Digital
Terrain Model. See Plate IX.
Digital European Cordless Telephone, DECT
emerging personal communications network standard in Europe,
employing CT-3 technology.
Digital Feature Analysis Data, DFAD
a federal government graphic data structure.
digital hierarchy
sequence of standardized increments for multiplexing digital
channels.
digital incremental plotter
digital device to draw line segments.
Digital Line Data, DLD
digital data recorded in the computer as a string of digital points.

Digital Line Graph, DLG

a group of federal government digital spatial data exchange formats for graphic products; defined in several iterations and enhanced versions; USGS employs DLG formats containing coordinates describing contour data and related boundary, drainage, and other features.

digital loopback

a diagnostic test forming a loop at the modem interface.

digital marginalia

part of a spatial data base containing information normally stored on the margin of a paper map, including information such as map accuracy, compilation date, last revision date, and so forth.

Digital Multiplexed Interface, DMI

in Local Area Networks, a voice and data PABX standard for T1 transmission, supported by AT&T.

Digital Number, DN

numerical value of a pixel--remote sensing term.

Digital Service Unit, DSU

peripheral enabling the connection of computers across digital lines supporting voice and data transmission.

Digital Serving Area, DSA

the combined geographical serving areas of a set of DDS serving offices, as specified in the appropriate tariff(s); the DDS office serving areas comprising a DSA are not necessarily contiguous, and a DSA may overlap state boundaries.

digital system

a system for discrete numerical information transmission normally using a binary code representation such as a relay operated or non-operated; a solid state device conducting or non-conducting; or a pulse's presence or absence.

digital tablet

device used to hold a map and to transmit to the computer, using a cursor, the location of electronically sensed coordinates of points on the map.

Digital Terrain Elevation Data, DTED

a federal government digital graphic data format for terrain elevation data.

Digital Terrain Model, DTM

a coordinatized three dimensional cartographic model (solid) of digital elevation data; often displayed as a grid (fishnet) draped over underlying topography. A generic term describing the digital cartographic representation of terrain elevations for ground positions at regularly spaced intervals. See Plate IX.

Digital to Analog converter, D/A
device that converts digital values to analogue voltages in order to manipulate the beam of electrons in a video display.
digitize
the input to the computer of those facilities placed on the land base during manual layout.
digitized data
any items recorded and stored in a computer as digital data.
digitizer
a computer peripheral used to convert analog quantities such as map information into discrete digital samples. This peripheral consists of a table and a cursor (mouse) with crosshairs and buttons used to record the locations of map features as x,y Cartesian coordinates in the attached computer. It can be a flatbed digitizer or a scanning digitizer.
digitizing
data entry process in which graphic images are changed into digital coordinates.
digitizing cursor
manual digitizing cursors, with attached small scanning windows are available as a compromise among various scanning options.
digitizing table, tablet
peripheral used to convert analog graphic data to digital format for further computer processing. Conductors in the table transmit electrical signals from a cursor to the computer. Older units, now obsolete, used microphones and sonic transmission to achieve such conversion.
digtizer accuracy
maximum error in x and y axis positions between true and recorded coordinates of a given point.
digraph
a graph with orientation assigned to the edges; often used to represent flows in a real-world network.
DIME
see Dual Independent Map Encoding.
dimension
measure of space filling; points have dimension zero, lines have dimension one, areas have dimension two, volumes have dimension three, and hyperobjects have higher integral dimension. All of these are objects of integral dimension; one can also calculate fractional dimensions (instead of Euclidean dimension) using fractal geometry. Euclidean dimensions offer a set of discrete values as possible measures of objects; fractional dimensions offer a continuous spectrum of measures. Fractional dimensions are useful for evaluating scale change and zooming in, and more generally, any process in which enlargement

or reduction of an image occurs. Currently, electronic maps tend to function only with regard to Euclidean dimension. Plate IX suggests a three dimensional object in two dimensions, as does Plate III.

dimensional stability

resistance of paper or film to curl or exhibit other dimensional distortion.

DIN

see Deutsche Industrial Norms.

DIP

see Dual In-line Pins.

direct access

data access mode in which a user can interact directly with the central processing unit of the computer.

direct positive

the positive image obtained by exposure in the camera with subsequent chemical treatment to develop and "reverse" the tones of the image.

directed graph

see digraph.

directory

a table cataloging the files in a given account.

Dirichlet region

geometric process of partitioning an area in such a way that points are sorted into regular tiles according to the minimum distance between them and a set of predetermined points. An arbitrary convex hexagon has a Dirichlet region contained within it that is a centrally symmetric hexagon. Tesselations of various sorts are useful in partitioning space into regular patterns that are useful for analytic purposes. Also, Thiessen polygons and Voronoi polygons. See star.

disc

see disk.

discrete

separated numerical entities, not capable of interpolation. Contrast with continuous. Discrete mathematics is that large grouping of mathematical subfields which studies the logical character of numbers (number theory), graphs (graph theory), and a host of other topics; the theoretical basis of electronic mapping rests on discrete mathematics. Plate XI illustrates a discrete use of color to show different speeds.

discrete access

in Local Area Networks an access method used in networks with a star topology. Each station has a separate connection, as contrasted with a shared connection.

discriminant
  algebraic tool sometimes used to partition a space of measurements.
disk, disc
  an electromagnetic storage device for digital data of all kinds. Permanently mounted discs are termed "hard" while removable discs or diskettes are termed "floppy" owing to the two different recording medium sub-strates.
disk cache
  memory set aside on a temporary basis to hold data in order to speed operation of the system.
disk drive
  a computer mass storage peripheral that records and retrieves digital data to and from a disk or diskette without loading data.
disk fragmentation
  see data fragmentation.
Disk Operating System, DOS
  a microcomputer operating system by IBM and by Microsoft Corporation. It is a set of programs that instructs a disk-based computer to manage the entire system, including peripherals--it is widely used in the microcomputer industry.
Disk Save and Compress, DSC
  a process that compresses data and space, resulting in available computer memory being accessible in a contiguous form.
disk space
  the amount of computer memory available for processing; some software systems require that this space be contiguous,   rather than randomly distributed.
diskette
  electronic storage device--floppy disks. These disks are typically reusable and are inserted into a disk drive by the user. Software is often sold on them; they are easily portable; and, they currently are commonly available in 5.25" and 3.5" sizes in a variety of styles--high density, double-sided double density; formatted or unformatted.
dispersion
  process in which rays of light of different wavelength deviate angularly by different amounts from each other. A glass prism disperses the rays of white light. Diffraction gratings also disperse light. Also see fiber dispersion.
dispersion-limited
  in optical fiber transmission, the limits placed on the optical pulse by dispersion at the end of the transmission line.

displacement
shift in the position of an image for any of various reasons including scale change.

display
graphic image of output data.

display board
computer circuit board used to translate display data into video signals for the CRT.

display resolution
level of detail available in an image; on a computer display screen it is expressed in terms of numbers of pixels covering the screen, as a matrix--width by height. Low resolution is 320 by 200 pixels; medium resolution is 640 by 400 pixels; and, high resolution is 1280 by 1040 pixels.

Displayable Attribute Processing, DAP
the process of graphically displaying the information in a data base.

dissolve lines
lines between adjacent areas that are merged to form larger areas particularly when adjacent areas represent the same attribute. The boundaries of the 36 tiles in Plate VI dissolve to form the image of the Mississippi drainage basin.

distance
often geographic computations of various sorts involve analysis based on distances. Vehicle routing using a GIS may invoke any of a variety of procedures, such as those from the Travelling Salesman Problem, and from various shortest path algorithms.

Distances may be simple Euclidean distances or they may be based on a metric that is less common. Manhattan distance, measured along grid lines is useful when number of turns is important; social distance, which may not be at all physical, may be important in understanding various real-world patterns. Distance measured on a sphere (great circle distance) is important in looking at global issues. The manner in which distance is measured should be considered at the outset of any geographical analysis so that the distance that best suits the problem might be selected; otherwise, inappropriate results might follow.

In Plate VIII distances in one layer are measured "as the crow flies" and in the backdrop are measured along streets. In Plate VII distances in both layers are measured along streets.

distortion
any shift in the position of an image on a photograph which alters the perspective characteristics of the photograph. In a telecommunications environment, an undesired change in signal between sender and receiver.

distortion elimination
    capability for removing distortion generated by data input procedures.
distributed architecture
    a Local Area Network using the bus or ring topology that uses shared communications.
distributed computing
    used to describe the idea of moving computing resources closer to the individual user.
distributed database
    database with components in different locations linked by telecommunications network; SQL may be used to distribute a database to many computers, creating a network of micro or minicomputers with vast computing capability.
distributed processing
    a strategy by which separate computers share work on a single application program, instead of concentrating all the word in a single location.
distributed system
    computer network in which data processing is done in a variety of locations on separate computers, instead of concentrated in a single large CPU.
Distributed-FeedBack laser, DFB
    a laser in which part of the output is fed to to the input using more than one propagation mode in the feedback path; it operates more efficiently when feedback and direct waves are in the same mode.
distribution block
    centralized telecommunications connection equipment where cross-connections are made.
distribution cable
    cable connecting customers to a feeder cable, serving area interface or similar device.
districting
    process of determining districts as an accumulation of distances from pre-selected nodes.
dither/dithering
    a printing process that introduces slight random displacement of dots to create an effect of continuous change along edges that would otherwise be viewed as discrete breaks. Dithering is primarily a form of smoothing. Dithering also has application in radar tracking.
DLC
    see Data Link Control.

DLD
   see Digital Line Data.
DLG
   see Digital Line Graph.
DMI
   see Digital Multiplexed Interface.
DML
   see Data Manipulation Language.
DMRS
   see Data Management Retrieval System.
DN
   see Digital Number.
DNIC
   see Data Network Identification Code.
documentation
   written specifications and explanation of how to use a computer program; indicates hardware support required to run the program, as well as various instructions on how to format the input data and how to use the program to create desired files.
DOD
   see Department Of Defense.
domain
   the domain of a mathematical function is the set of input values on which the function, or mathematical mapping, is to act, producing a set of output values in a set (called the range). In a similar vein, in a GIS setting, the concept domain refers to a set of (indecomposable) values from which an attribute takes on values.
dominant wavelength
   monochrome light that will match the target color when mixed appropriately with the illuminant light.
dongal/dongle
   hardware lock used to restrict access to software; see hardware key.
dopant
   a material combined, in any of a variety of ways, to another material to produce a set of desired characteristics.

doped-silica cladded fiber
an optical fiber of doped silica core and doped silica cladding, produced by the Chemical Vapor Deposition process.

doped-silica graded fiber
an optical fiber with a silica core. The doping varies to produce a decreasing refractive index from the core outward. The need for cladding is eliminated--compare these two doped-silica fibers to the ideas of discrete and continuous.

DOS

see Disk Operating System.

dot distribution map

map using dots to show density variations; each dot represents a constant amount (often expressed in numbers or percentages) of the attribute mapped. Thus, a dot map might show state population with dots spread across each county with each dot representing 50,000 people, or perhaps with each dot representing 1% of the population of the state. In using dot maps as analytic tools, one must be careful to distinguish whether the individual dots represent numbers or percentages, and how the dots are scattered within the smallest areal unit for which data is collected.

dot matrix plotter

a plotter in which dots printed on the paper make up the image of the map; the spacing of the dots is determined by the spacing between wire points in the printing head (100 to 400 per inch).

dot matrix printer

see matrix printer

dot pitch

distance between phosphor dots on adjacent lines of a CRT; to a point, the smaller the dot pitch, the better the image.

double line drain

digitized gradient that is flat in width and the slope is greater than or equal to 0.

double precision, double coordinate precision

uses double the number of bits in a word to express number of significant figures in calculations; in a 32 bit computer, when a double word of 64 bits represents real numbers, the resulting calculations are precise to 8 significant digits; double coordinate precision refers to the same idea for pairs of numbers. Double-precision images can achieve an accuracy of less than one meter at global resolution.

double-digitized polygons

when polygons are recorded as closed loops of coordinates, the first and last coordinates are the same; therefore, boundaries of adjacent polygons are digitized twice.

Double-Sided Double Density, DSDD

a type of magnetic storage floppy disk.

down loading

see downline loading.

downline loading

the process of sending files, software, or other electronic matter from a central computer to individual stations.

downloadable font
font that is installed in software and sent to the printer prior to printing.

DP
Data Processing; manipulation of data using a computer.

DPSK
see Differential Phase Shift Keying.

drag
motion with a mouse cursor in which an item is selected and pulled across the screen by holding down a mouse button.

drain line
a line feature depicting drainage flow. See Plate IX.

drape
in a GIS context to cover one surface with another--as to drape a road network over an underlying visual display of terrain. In Plate I, color is draped over a grid/fish net (right side) to show differences in accessibility.

Drawing eXchange Format, DXF
a graphic file and data spatial exchange standard, developed by Autodesk, Inc.

drawing sizes
A-size 8 1/2" x 11"; B-size 11" x 17"; C-size 17" x 22"; D-size 22" x 34"; E-size 34" x 44". Corresponding metric sizes (of somewhat different dimensions) are A0, A1, A2, A3, and A4.

DRC
see Design Rules Checker.

drift
the lateral shift or a displacement of an aircraft from its course, primarily because of the action of wind. This concept has similar expression in Kriging, in which drift refers to displacement of a regionalized variable in one direction.

driver
software to control a port or a peripheral--a printer driver, for example.

drop
node on a multipoint circuit.

drop shadow
a background shadow used to highlight a graphic; suggests that it stands out from the page. Drop shadows are used to set off the map projections displayed in this book.

drop wire
one or more pairs of relatively short, insulated wires that run from a distribution terminal to the protector on a customer's premises.

drop-line; drop-arc

removal of boundaries between map units bearing the same identifying label.

drum plotter

plotter with a rotating cylinder used as a drawing surface; thus, output of images of arbitrary length, but of maximum width the axial length of the cylinder, is possible.

drum scanner

device for automatic conversion of maps to digital form.

DS-0

Digital Signal - a T1 interface system-based communication channel of 64 kbps; when used for data, 8 kbps are used for in-band signaling and 56 kbps are used for data traffic.

DS-1

single circuit carrier consisting of 24 DS-0 channels; bit rate = 1.544 Mbps.

DS-2

carrier consisting of 96 DS-0 channels (equal to 4 DS-1s); bit rate = 6.312 Mbps.

DS-3

carrier consisting of 672 DS-0 channels (equal to 28 DS-1s); bit rate = 44.736 Mbps; DS-3 is the practical limit for short-haul microwave.

DS-4

carrier consisting of 4032 DS-0 channels (equal to 168 DS-1s); current upper limit for carrier.

DS-5

theoretical carrier for fiber optics.

DSA

see Digital Serving Area.

DSB

a terminal that is used with the buried plant.

DSBS

a splice closure used in buried plant.

DSC

see Disk Save and Compress.

DSDD

see Double Sided Double Density.

DSU

see Data Service Unit

DSX-0

digital cross-connect used to interconnect equipment at the DS-0 level.

DSX-0A

(STC X-conn) - the digital cross-connect at a DDS hub office where individual customer circuits are routed and test access is available.

DSX-0B

(Multiplex X-conn) - the digital cross-connect at a DDS hub office used to connect T1DM and T1WB4 ports with SRDM's.

DSX-1,2,3

digital cross-connect used to interconnect equipment, provide patch capability, and provide test access at the DS-1, DS-2, or DS-3 level, respectively.

DTE

see Data Terminal Equipment.

DTED

see Digital Terrain Elevation Data.

DTM

see Digital Terrain Model.

DTMF

see Dual-Tone Multiple-Frequency.

Dual In-line Pins, DIP

describes the pin arrangement on an integrated circuit or an electric switch.

Dual Independent Map Encoding, DIME

a U. S. Census Bureau data coding system, based on Standard Metropolitan Statistical Areas of the U.S.A., that combines x,y coordinate (location) encoding and topological (direction) encoding. When street segments are assigned x,y coordinates as well, the resulting geographic file is called the Geographic Base File/Dual Independent Map Encoding (GBF/DIME) file. One issue to watch for in such files is whether or not street addresses appear on the correct side of the street, even when the actual numbering pattern (from field evidence) is not one of the usual numbering patterns (such as odd on one side of the street and even on the other). To distinguish which side of a line a particular location is on requires that the Jordan Curve Theorem be part of the underlying topological basis on which the computer makes decisions.

Dual-Tone Multiple-Frequency, DTMF

audio signaling frequencies on telephones with Touch-Tone service.

dumb terminal

a terminal that uses no communications protocol, has no local processing power, and is therefore reliant on the processing power of another device. Teletypes and certain workstations are typical of dumb terminals.

duobinary coding

a signal design technique that codes and shapes binary data signals into a special waveform, characterized by three voltage levels; this process results in a two-to-one bandwidth compression, hence twice the data capacity for a given bandwidth; duobinary coding also permits the detection of errors without the addition of error-checking bits.

DXF

see Drawing eXchange Format.

E e

Earth Observation SATellite Company, EOSAT

owner of Landsat satellites since 1986.

Earth Resources Technology Satellite-1, ERTS-1

a mapping and remote sensing satellite launched in 1972; later renamed LANDSAT.

earth station

the ground equipment, including a dish and its associated electronics, used to transmit and/or receive satellite communication signals.

easement

right-of-way over another's ground; may be a public or private restriction on land use. On Plate V, the property lines go to the center of the street, so that the street easement is visible.

easting

measurement of Cartesian coordinate distance east a location is from a north-south meridian or other reference line; false easting is an adjustment constant to eliminate negative numbers.

EBCDIC

see Extended Binary Coded Decimal Interchange Code.

ECC

see Error Correction Code.

echo

the return to the sender of transmitted data.

echo suppressor

equipment that blocks receiving capability while transmission is occurring.

echoplex

a test for data integrity involving the return of particular transmitted characters to the sender for verification.

edge

link joining two nodes of a graph. In Plate VIII, the central node within a spider (star) is joined to each of n other nodes along n edges.

edge emitting LED

a light-emitting diode; more efficient for intensity of output and coupling of optical fibers than is a surface-emitting LED.

edge enhancement

use of quantitative tools to highlight selected pixels to enhance image edges.

edge match(ing)

elimination of discrepancies at the edges of adjacent map sheets to create a continuous image; thus, every line that leaves the Plat Border (R/W, Water, Boundary, Cable, Conduit) must go to the adjacent map and match exactly, no exceptions.

The process of joining adjacent digital map sheets and insuring connectivity and continuity between sheets by adjusting features to produce agreement of values along the boundary. See Plate II; the edges of three images match.

edit

to find and correct errors in text or images.

edit and display on input

in a GIS, continuous display and editing of input data during period when digitizing is ongoing.

edit and display on output

in a GIS, the capability to preview and edit displays prior to the printing of hard copy.

edit verification

the comparison of the source documents to the mechanized PLRs to assure all items are transferred correctly.

EDM

see Electronic Distance Measuring device.

EGA

see Enhanced Graphics Adapter.

EIA

see Electronics Industries Association.

EISA

see Extended Industry Standard Architecture.

EL

see Electro-Luminescence.

ELAS

NASA-developed (Stennis Space Center) public domain image processing system.

elastic box

box on an image display used to select areas for zoom, clip, and similar actions, on a map; box can be resized or moved with a mouse.

elastic circle
  see elastic box.

Electro-Luminescence, EL
  a computer screen display technology (cf. LCD).

ElectroMagnetic Radiation, EMR
  energy propagated as an advancing wave of interaction between electric and magnetic fields. Light and radio waves are examples.

electromagnetic spectrum
  remote sensing devices record electromagnetic bands from this spectrum, that covers radiation wavelengths from gamma rays of length 0.001 angstrom to long waves of length over 1 million kilometers; most remotely sensed lengths fall into the visible or infrared category, currently. Words commonly used to describe various ranges of wavelengths include: ultraviolet, visible, infrared, and microwave radiation.

Electronic Distance Measuring device, EDM
  an electronic replacement for the traditional surveyor's optical transit.

Electronics Industries Association, EIA
  a U.S.A. standards organization.

electronic journal, e-journal
  journal transmitted electronically over a public or private communications network; *Flora On-Line* and *Solstice* are two examples of electronic journals.

electronic mail, e-mail
  messages transmitted electronically over a public or private communications network; Bitnet and Internet are two examples of public networks.

electrostatic plotter/printer
  computer peripheral used to produce raster format hard copy of graphic images by placing electrical charges on paper so that dark toner powder will stick to these charged dots. A dot-matrix printer is an example of such a device.

element
  a discrete item of information (analytical, graphic, or textual) or input to a data base; an element may be a data point, an attribute, or a group of related attributes; generally, a member of an abstract, mathematical set. One graphic element of Plate VII is a bus route formed from nodes and edges.

elevation polygons
  regions of close-to-uniform elevation. See Plate IX

elevation post

a data point (x,y,z) related to a defined digital data base and map projection that is referred to as "z" only (elevation). See Plate IX-- heights are the z dimension.

ellipsoid

ellipse of revolution that may be generated by rotating an ellipse through 360 degrees about its major or minor axis; when used to represent the earth, the rotation about the minor axis gives a closer approximation. Standard ellipsoids, made from careful surface measurements, to represent the surface of the earth have been defined at various points in history; in 1866 by Clarke; in 1924 one was adopted describing the flattening of the earth as 1 part in 297; currently, satellite data is used to construct ellipsoid models that tie to physical characteristics of the earth's mass.

embedded SQL

feature in a relational database management system to write programs in host computer languages using blocks of SQL statements.

emissivity

given constant temperature, the ratio of radiation from a surface to radiation from a blackbody.

EML

see Expected Measured Loss.

empty slot ring

a local area network with a ring topology in which a free packet circulates through every station.

EMR

see ElectroMagnetic Radiation.

EMS

see Expanded Memory Specification.

encapsulation

covering cable pairs in a splice or at a cable end with a jelly-like compound that insulates and protects cable from moisture.

enclosing rectangle

smallest rectangle to serve as an envelope containing an element of an image.

encode [1]

process of converting a signal from one format to another (e.g., a signal's pulse amplitude into a binary code or machine readable format).

encode [2]

to symbolize information.

encryption

file encryption is the process of applying an algorithm to data to concealing its content; decryption is the procedure that reverses the operation and converts encoded information back to its original content.

End Of File, EOF

computer code used to end data set; a special marker, placed at the end of data in a storage file.

End of TeXt, ETX, EOT

control character used to mark the End of TeXt.

end office

a local central office arranged for terminating subscriber lines and provided with trunks for establishing connections to and from other central offices.

end node

the endpoint of an arc that was digitized last.

end points

the points (nodes) marking the two ends of a line string.

energized pairs

pairs that are connected to the main distributing frame.

engineering map

often a large scale map showing information particularly pertinent to engineering projects: a map showing roads, infrastructure, property lines and utility connections might be used to display spatial elements of a municipal road construction project. See Plate V.

Enhanced Graphics Adapter, EGA

a type of microcomputer graphics display and/or monitor with horizontal scanning frequency ranging of 21.85 KHz. This adapter provides resolution superior to CGS but is limited to 64 colors.

Enhanced Small Device Interface, ESDI

a device-level computer peripheral interface.

ENQ

see, enquiry

ENQ/ACK protocol

a Hewlett-Packard communications protocol.

ENQuiry, ENQ

a control character used to request identification or status. In ASCII code, it is "control E."

entity

a logical structure in the data base that is represented with a name and number.

entity subtype/supertype

identical with the mathematical notion of subset and superset; a set A is a subset of a set B if and only if for every element a in A, a is also in B; in that event, B is said to be the superset of A.

enumeration unit

areal unit for the collection of quantitative data.

EO cartridge

see Erasable Optical cartridge.

EO drive

see Erasable Optical drive; also, CD-MO.

EOF

see End Of File.

EOSAT

see Earth Observation SATellite Company.

EOT

see End Of Text.

EPPL7

see Enhancement of Environmental Planning and Programming Language.

EPROM

see Erasable Programmable Read-Only Memory.

equal area map projection

equal area projections are useful for small scale maps; they are such that a unit square represents the same extent of land/water independent of its position on the map. The equal area character comes at the expense of conformality. These projections are thus useful for comparing extent of geographic areas--such as broad expanses of forested lands, the world's ocean/fishing surface, or climatic circulation patterns. There are a great many attractive projections from which to choose: choice depends on the size of the area considered and on the distribution of angular deformation.

Albers' equal area conic projection (Figures 1 to 4) has no angular deformation along its two standard parallels; it is a good choice for middle latitude maps of relatively large east-west extent. On Lambert's equal area projection (Figures 11 and 12), distortion is symmetric about the center; thus, when the projection center is the center of a continent, this choice is quite reasonable for display at the continental scale.

On a sinusoidal projection (Figure 20) there is no angular distortion along the Equator or the central meridian, and the rest of the meridians appear to be evenly spaced (but are not); thus, this projection is useful for displaying landmasses of considerable latitudinal extent, such as Africa.

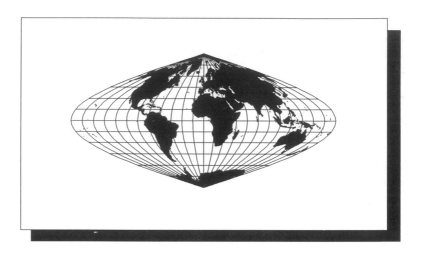

Figure 20. Sinusoidal projection.

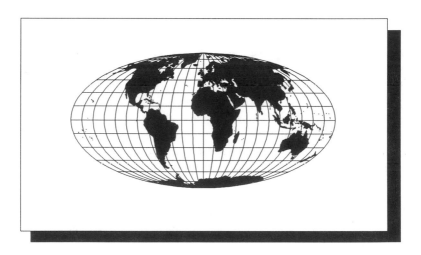

Figure 21. Mollweide projection.

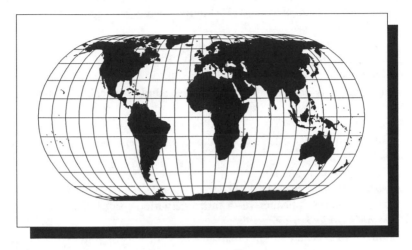

Figure 22. Eckert IV projection.

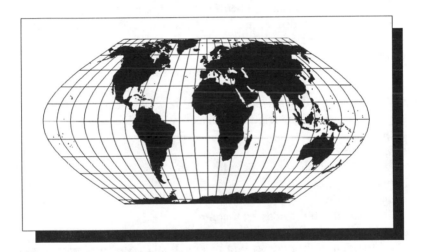

Figure 23. Eckert VI projection.

Mollweide's equal area projection (Figure 21) has a blunter shape than the sinusoidal; it is particularly well-suited to portraying middle latitude distributions. Allen Philbrick has combined the advantages of both the sinusoidal and the Mollweide, to create an interrupted "sinu-Mollweide" projection.

Eckert's projections have characteristics similar to those of a Mollweide projection, except that compression of landmasses at the poles is not as severe as in the Mollweide. Two of Eckert's projections in common use are numbered IV (Figure 22) and VI (Figure 23).

equalization

compensation for telephone line distortions.

equalized histogram

statistical histogram in which the distribution has been mathematically balanced to even out the number of cells of each data value.

equalizer

device used by modems to compensate for telephone line distortions.

equinox

see circle of illumination.

Erasable Optical cartridge, EO cartridge

see Erasable Optical drive--a cartridge with the same characteristics.

Erasable Optical drive

drive to hold erasable optical disks; high capacity storage device, similar to a WORM disk, holding up to 500 megabytes per side; however, EO disks can be re-used while WORM disks cannot.

Erasable Programmable Read-Only Memory, EPROM

a nonvolatile semiconductor PROM whose content can be cleared to accept new information.

error

difference between actual and computed values.

error coding

process of checking existing plats for errors and identifying the errors for the reconciliation team to correct. The first stage in record preparation.

error control

a procedure that provides for both error detection and error correction.

error controller

equipment that provides error control.

error correction

restoration of data integrity in received data. Such restoration may come about by manipulating the received data according to well-established mathematical principles from coding theory or it may come about by requesting retransmission of the data from the source.

Error Correction Code, ECC

a means of reducing hard error rates in storage and transmission media.

error detection

discovery of errors in transmitted data by the receiver. Such discovery may come about using various techniques from coding theory. Detection does not imply correction, which is a separate process.

error message

message displayed by the computer when it detects a problem in a program during its compilation, interpretation, or execution.

error rate

a measure of data integrity expressed as the fraction of bits which are flawed.

error rating

the process of assessing the type and number of errors and applying a weighted value to each to arrive at the percentage of errors in the wire center.

error reconciliation

the process of research and correction of missing or erroneous data on the original map data sources. The second stage in record preparation.

ERTS-1

see Earth Resources Technology Satellite-1.

ESDI

see Enhanced Small Device Interface.

essential facilities

in contrast to additional facilities, those which are available on all networks in packet-switched networks.

Ethernet

a network protocol defining a specific implementation of the Physical and Data Link Layers in the OSI model (IEEE 802.3). Ethernet is a local area network, using a bus topology, that provides reliable high-speed communications (maximum of 10 million bits per second) in a limited geographic area, such as an office complex or a university campus. See also thin Ethernet; also referred to as "thinline."

Ethernet circuit

ethernet circuits provide for a multi-access connection between a number of nodes on the same circuit; an Ethernet circuit is a path to many nodes; each node on a single Ethernet circuit is considered as being adjacent to every other node on the circuit and equally accessible.

ETX

see End of TeXt.

Euclidean geometry

often referred to as plane geometry, the geometry of Euclid that studies relations of points, lines, and planes using a systematic logical structure. So-called non-Euclidean geometries are properly those that do not obey Euclid's fifth postulate, that given a line m and a point P not on m there is exactly one line through P that is parallel to m. Elliptic geometry occurs when there are no parallels, and hyperbolic geometry when there are more than one parallel. Often, however, non-Euclidean is used in a variety of other ways which presumably are clarified by the context in which they are used.

exchange

a set of central offices used as an administrative center of communications serves for a common carrier within a specific geographic territory.

exchange area

an area within which there is a single uniform set of charges for telephone service. An exchange area may be served by a number of central offices. A call between any two points within an exchange area is a local call.

exchange cable

cable between the central office and a subscriber.

executable code

see machine code.

existing allocation area

an allocation area that reflects the existing configuration of the network. Existing allocation areas are used to monitor problems and to establish the base for evolution to the ultimate design. An existing allocation area is a group of areas fed by individual lateral cables from the same feeder cable. The area is selected to produce a minimum of multipling with other existing allocation areas. The area should lie within boundaries that resemble the boundaries of the associated ultimate allocation area. As existing allocation areas become ultimate allocation areas, multipling between allocation areas will disappear. On the Cover Plate the spiders (stars) show the assignment of students to existing schools.

existing splice

the point in a cable where a physical joining of pairs has taken place.

expanded memory

memory beyond conventional memory; a standard for enlarging a computer's memory that pre-dates extended memory.

expanded memory emulator

program to convert extended into expanded memory when needed for various software applications.

Expanded Memory Specification, EMS
a method of exceeding the 1 megabyte main memory constraint of DOS.

expandor
device to reverse analog compression.

expansion board
a plug-in circuit board that increases the capabilities of a microcomputer.

expansion slot
socket for optional circuit card on the motherboard of a microcomputer.

Expected Measured Loss, EML
the expected reading, in db, of a TMS connected to a prescribed test point at one end of a trunk when a sending power of a specified value is applied to a prescribed test point at the other end of the trunk.

expert system
computer system that attempts to simulate, using rule-based logic, human intelligence. See artificial intelligence; neural network.

explicit access
a method of sharing access in a Local Area Network in which each station is guaranteed a turn at access but must wait in line for that turn. In contrast with contended access.

export
transfer of data from one system to another.

Extended Binary Coded Decimal Interchange Code, EBCDIC
an 8-bit character code, principally used in IBM equipment, that allows for 256 different bit patterns assigned to standard QWERTY keyboard characters.

eXtended Graphics Adapter, XGA
includes VGA and resolutions up to 1024 by 728.

Extended Industry Standard Architecture, EISA
a 32 bit micro-computer bus design/architecture.

extended memory
memory to enlarge conventional memory; requires a memory manager, that keeps different applications from using the same part of the extended memory at once, so that DOS applications can access this memory. As contrasted with expanded memory; EMS vs XMS; diagrams of memory stacks and related acronyms involving memory are available in many manuals that accompany software (depending on the memory requirements of the software).

extract
process of transferring data out of one system.

extrapolation
extension of results beyond the set of points actually sampled; when default curves or straight lines are used to perform such extension, the user needs to consider very carefully whether the results are meaningful and whether or not a choice superior to the default curve might have been used to get an extension that fits better with reality. This is quite important, since extrapolated data is often used as evidence on which to allocate funds in accordance with various policies that have a future component.

F f

4GL
see Fourth-Generation Language.
facet
a digital map "sheet," in some systems equivalent to a design file.
Facilities Management
see, Automated Mapping/Facilities Management.
facility
a capability offered by electronic equipment or software.
facility network
infrastructure networks, including water, sewer, gas, cable TV, electric, and telephone transmission lines; these are often tracked using GIS or AM/FM. See Plate V.
facility splice
a physical connection of two or more cables that involve working pairs.
factor analysis
statistical tool to reduce the size of a set of variables. Some of the variables are expressed in terms of the others, so that each variable in the reduced set is a function of some of the original variables.
false color film
registers green, red and infrared reflected radiation, but not blue. See Plate III.
fault tolerant
pertaining to software or systems which execute properly even though parts may fail.
FCC
see Federal Communications Commission.
FCS
see Frame Check Sequence.
FDDI
see Fiber Distributed Data Interface.

FDHD

see Floppy Dive High Density.

FDM

see Frequency Division Multiplex.

FDMA

see Frequency Division Multiple Access.

feature

in a mapping context a feature is represented as a contiguous area (which may have holes; topological genus is not an invariant of a feature) of related content that is different from surrounding areas. Thus, "land" is a feature that includes islands and therefore the lakes containing islands are holes in the land feature. Plate VI shows rivers as features different from surrounding territory; indeed, there are "holes" in the red network in which only the underlying blue network is visible.

feature attribute table

arc or polygon attribute table in which coverage attribute information is stored.

feature code

alphanumeric label for a feature.

feature selection by attribute

logical process of selecting a subset of features from a coverage; only features whose attributes fall within the constraints are selected.

feature separation

preparation of separate image layers or objects for selected features in a data base. In Plate VII, the bus routes are in one layer and the road grid in another.

Federal Communications Commission, FCC

this U.S.A. commission is composed of seven individuals appointed by the President of the U.S. to regulate all interstate communications systems and all international communications systems that begin or end in the U.S. This commission was established by the federal Communications Act of 1934.

feature user-ID

user-assigned number given to each feature.

Federal Geodetic Control Committee, FGCC

a standards committee, concerned with geodetic control, of the federal government of the U.S.A.; this committee has published many documents concerning accuracy levels.

Federal Information Processing Standards, FIPS

U.S. Federal Government source for information processing standards. FIPS #173 establishes standards for the formatting and exchange of spatial data--based on reports of the Moellering Committee.   FIPS #5 is ZIP Codes.

FIPS provides a government standard state and country identification code; standard approved for use by U.S. governmental agencies.   FIPS #173 is the recently adopted Spatial Data Transfer Standard which includes standardized definitions for a variety of digital mapping terms; these definitions address federal requirements for accuracy and may or may not serve a practical need for everyday users. The federal effort at standardization is the authority.

feeder cable

one of several large cables leaving a central office that provides physical connection between the central office and distribution cables. Feeder cable is usually placed in underground conduit with access at manholes. These might typically be displayed on a digital map such as the one in Plate V.

fence

a "border" used in board digitizing to delimit a set of elements within an area for treatment as a single unit for graphic operations such as clipping or transformation.

ferrous

composed of and/or containing iron; a ferrous metal exhibits magnetic characteristics as opposed to a non-ferrous metal, such as aluminum.

FET

see Field-Effect Transistor.

FF

see Form Feed.

FGCC

see Federal Geodetic Control Committee.

fiber buffer

material surrounding an optical fiber. It isolates and protects the fiber mechanically.

fiber dispersion

increase in the width of a transmitted electromagnetic energy pulse as it moves along a fiber. This increase is caused by material dispersion which is the result of the frequency dependence of the refractive index. Modal dispersion is caused by different group velocities of different modes. Waveguide dispersion is caused by the frequency dependence of the propagation constant of that mode.

Fiber Distributed Data Interface, FDDI

a media access control-level protocol with token ring architecture and 100 Mbps throughput.

fiber optic cable
transparent glass or plastic cable that carries data sent via light impulses generated by lasers.   Each slender strand of glass or plastic provides a path for light rays to act as a carrier of the transmission. Incoherent optical fibers transmit light, but not images; coherent optical fibers can transmit an image through small, clad optical fibers.

fiber optic transmission
transmission of data through a fiber optic cable; see fiber optic cable.

fiducial marks
index marks, usually four, which are rigidly connected with the camera lens through the camera body and which form images on the negative and usually define the principal point of the photograph.

field
within a record, a group of bits that contains a specific category of information.

field check
verification of map accuracy by checking the actual position of features and attributes on the Earth.

Field-Effect Transistor, FET
a transistor controlled by voltage rather than current; the flow of working current through a semiconductor channel is switched and regulated by the effect of an electric field exerted by electric charge in a region close to the channel called the gate; also called unipolar transistor; a FET has either P-channel or N-channel construction.

Field Of View, FOV
in remote sensing, the dihedral (solid) angle through which a sensor can sense radiation.

Field-Programmable Gate Array, FPGA
an array of gates on a chip whose interconnections can be arranged electronically by the user.

field reporting code
two part, alphanumeric code used to identify type of plant and work operation; same as plant account code.

field splicing
joining of ends of optical cable, without the use of detachable connectors, in such a way that all internal elements of the cable function in a continuous manner.

FIGS
see FIGures Shift.

FIGures Shift, FIGS
a shift that enables the printing of numbers and symbols, based in Baudot Code.

file

a collection of records (data) treated as a unit.

file format

pre-designed layout for data in a file.

file name

identifying designation for an individual file.

file server

network computer that stores and manages files in shared or private subdirectories.

file server protocol

a communications protocol in a Local Area Network. It permits applications programs to share files.

file specification

a unique name for a file on a mass storage medium; it identifies such characteristics as device, directory name, file name, and version number under which a file is stored.

file structure

arrangement of entries in a file.

File Transfer, Access and Management, FTAM

OSI standard for file exchange.

File Transfer Protocol, FTP

an ARPANET service that provides a family of commands for performing file transfer and directory operations over multi-vendor networks.

filename extension

typically, a dot (.) and three to eight letter code suffix for computer filenames used to indicate the status of the file.

fill pattern

pattern used in mapping to fill in empty regions on the map that represent some real-world content. The pattern may be abstract, as shading in various tones of gray, or it may be a repeated concrete image, as blades of grass for a swamp or palm trees along a tropical coastal beach. In Plate II, abstract pattern fills areas of the lower right side.

film recorder

output peripheral that produces slides or prints on photographic film.

filter

an electronic device designed to allow desired signals of several frequency bands, or desired spatial patterns, to pass through and to block others that are not desired.

filtering
>process of using a filter to enhance the quality of an image by smoothing boundaries and similar tasks.

final post
>items of plant posted to the PLRs after authorization is completed.

FIPS
>see Federal Information Processing Standard.

firmware
>software stored permanently or semipermanently in some form of Read-Only Memory.

fix
>to determine one's position on earth using, for example, a GPS receiver to find latitude and longitude coordinates.

fixed count
>the permanent connection of a group of cable pairs to the binding posts of a connecting block or terminal block at a terminal location.

fixed count terminal
>a distribution point at which access is only available to those pairs as designated by the engineer. For example, in a 25 pair terminal one would have access only to those 25 pairs assigned by the engineer to that terminal regardless of the number of pairs in the cable.

fixed disk
>see disk.

fixed length record format
>a file format in which all records have the same length.

flap
>see overlay.

flat file
>a two-dimensional array used as an interchange format for (digital) data.

flatbed plotter
>peripheral with flat drawing surface on which to draw a graphic image on paper using a list of point coordinates and pen codes from the attached software; may be done using a variety of colors. Flatbed plots are fixed by the size of the bed; roller plotters have more length capability.

flight altitude
>the vertical distance above a given datum, usually mean sea level, of an aircraft in flight or during a specified portion of a flight; in aerial photography, when the datum is mean ground level of the area being photographed, this distance is called flight height or sometimes absolute altitude.

flight line
a line drawn on a map or chart to represent the track of an aircraft.

flight plan
specified information relating to the intended flight of an aircraft that is filed orally or in writing with an air traffic control facility.

floating point
instead of a fixed position decimal, a "floating" position is sometimes used to represent numbers to improve CPU capability to calculate using real numbers.

floating point board
similar to a math co-processor chip, a printed circuit board designed to speed processing in a CPU.

floppy diskette
see disc/disk.

Floppy Dive High Density, FDHD
a type of magnetic storage floppy disk.

flow accumulation
the number of cells that drain into a pre-selected cell in a GIS raster watershed analysis.

flow control
procedure to control the transfer of messages so that, for example, there is no loss of data as a printer serving a PC approaches its buffer capacity.

flow direction
in a GIS raster watershed analysis, each cell of the raster is assumed to flow into one of its eight neighbors with the direction of flow expressed in degrees clockwise from left--the left neighbor is 0, the upper left corner neighbor is 45, the neighbor on top is 90, the upper right neighbor is 135, the right neighbor is 180, the lower right neighbor is 225, the lower neighbor is 270, and the lower left neighbor is 315. See Plate IX.

flow path
in a GIS raster watershed analysis, the drainage path, beginning at a selected point, the flow path seed, and continuing to the outlet. See Plate IX.

flow simulation
dynamic simulation of movement of a set of entities through a network; buses through a street network, platelets through a blood stream, freight cars through a railroad network.

FM
see Frequency Modulation; see Facilities Management.

focal length
   the distance measured along the optical axis from the optical center of the lens to the plane of critical focus of a very distant object.

font
   a style of type (and by extension, any symbol).

footlamberts
   a unit of measurement used for representing the brightness (image intensity) of computer terminals.

footprint
   general word suggesting a surface area in various contexts: e.g. desk area occupied by a computer or area on the Earth being sensed by a remote sensing device.

foreign
   any item of plant not owned by the telephone or utility company.

form 4962MR
   detail of outside plant units added, removed, or adjusted under a specific outside plant authority or field check.

Form Feed, FF
   printer control character (ASCII or EBCDIC) used to skip to the top of the next form (page).

format
   preparation of media with basic locational information required to use the media for data storage; physical organization of a data set.

FORmula TRANslator, FORTRAN
   FORTRAN is one of the earliest high-level programming languages. Short for FORmula TRANslator, it was designed particularly for scientific applications.

forth
   a high level fourth generation computer programming language.

FORTRAN
   see FORmula TRANslator.

four-wire circuit
   a two-way communications circuit using two paths, arranged so signals are transmitted in one direction only on one path, and in the other direction on the other path; the transmission cable may or may not use four wires, since the same ground wire may serve both directions.

Fourier analysis
   mathematical technique which can be applied to remove unwanted spatial frequency components, such as lines introduced by a faulty collection device, from an image

fourth-generation language, 4GL
   a computer programming language based on nonstructured language close to natural language; this high-level language must be translated

first into source code and then into machine code to be executed by a computer; used to customize database management applications.

fox message

test message using all letters on the keyboard--based on "The quick brown fox..." often used as a training exercise for beginning typists.

FPGA

see Field-Programmable Gate Array.

fractal

a geometric object of fractional dimension; one in which some amount of space is filled between usual integral dimensions. When fractals arise from zooming in and repeating a self-similar image at different scales, there is an obvious connection to GIS--whether this phenomenon is observed in coastlines, soil samples, or elsewhere.

frame [1]

time required to sample and encode 24 channels including signaling and framing information into a binary code pulse train of 193 time slots; a frame is 125 microseconds long--a transmission block.

frame [2]

a single video image consisting of two interlaced fields.

Frame Check Sequence, FCS

field used for error detection in bit-oriented communications protocols.

frame grabber

video device that freezes and digitizes a single video frame.

frame space

part of display board that determines which part of the CRT contains an image; it can be larger than the CRT display area.

framing

process of inserting control bits in a signal to identify channels.

framing detector

identifies incoming framing pulses for proper receiver channel timing alignment with the transmitter information.

FRC

Field Reporting Code (see PAC).

freeze-frame video

a device that transmits and/or receives still video pictures over a narrowband or mediumband telecommunications channel; may refer specifically to a still frame video unit that "grabs" the image from the camera or other video source and "freezes" it in a fraction of a second.

frequency [1]

the number of times a periodic action occurs in a unit of time.

frequency [2]

the number of cycles that an electric current completes in 1 second.

frequency curve

a graph characterizing value of the observation as a function of number of items observed.

Frequency Division Multiple Access, FDMA

radio transmission method allowing multiple users to gain access to a set of radio frequency bands without interference.

Frequency Division Multiplex, FDM

in optical communications, a multiplexing method that separates the composite bandwidth into channels. Each channel is then assigned a specific range of frequencies. One also encounters wavelength division multiplex (WDM) which involves the use of several distinct optical sources (lasers), each having a distinct center frequency; FDM may be used with any or all of those distinct sources.

Frequency Modulation, FM

one way to add information to a sine wave carrier signal; other ways include Amplitude Modulation and Phase Modulation.

frequency range

on a CRT, the upper and lower frequency limits that can be used.

Frequency Shift Keying, FSK

a commonly used modulation technique for sending serial binary data over communication links in which one frequency carries marks and another carries spaces.

frequency spectrum

array of frequency components in a given pulse, sound wave, or other recurring phenomenon.

frequency transform

procedure to partition an image into its fundamental spatial frequency components. These units may then be subjected to a variety of analyses.

Fresnel Reflection Loss

loss of power at a surface interface due to the reflection of incident power. Same as Fresnel reflectance loss.

from-node

the endpoint of an arc that was digitized first.

FSK

see Frequency Shift Keying.

FTAM

see File Transfer, Access, and Management.

FTP

see File Transfer Protocol.

full duplex

a facility that permits transmission in both directions concurrently.

function
   mathematical association of elements from one set (range) with each element of another set (domain). Thus, y=f(x) represents a function in which the x-values are the inputs (independent variable) in the domain to which the rule f is applied, yielding the outputs of y (dependent variable) in the range. To each input there corresponds exactly one output; if more than one output corresponds to each input, then the association is not a function, but a relation. Thus, y=x is a function; y=(+/-)x is only a relation. All functions are also relations; most relations are not functions.

functionality
   capability of a system to perform a particular operation.

fundamental frequency
   a base frequency as opposed to harmonics; in the U. S. 60 hertz is the fundamental frequency on the power network.

fusion splicing
   the joining together of two solid transmission media in optical transmission systems in such a way that no reflection or refraction occurs at the point of the joining.

fuzzy tolerance
   method of removing coordinates within a certain minimum distance of given nodes; a way to remove redundancy, retaining only those nodes critical to the analysis; one application of fuzzy logic.

G g

GaAs
   see Gallium Arsenide.

gain-bandwidth product (of an APD)
   the gain times the frequency of measurement when the device is biased for maximum obtainable gain.

Gallium Arsenide, GaAs [1]
   substrate for opto-electronic integrated circuits.

Gallium Arsenide, GaAs [2]
   dopant for optical fiber.

gamma rays
   very short wavelengths (10 to the (-6)) of the electromagnetic spectrum.

gantry
   the movable frame on a flat bed plotter that holds pens, pencils, or scribing tools in place over the drawing surface.

gap

distance between lines on a digitized map, often the result of a digitizing error.

garbage

corrupted data; see GIGO.

Garbage In, Garbage Out, GIGO

a familiar phrase, usually applied to data; however, the user of a GIS should also be wary--lack of thought or knowledge can lead to absurd visual output.

gate

an electronic switch; a circuit having several inputs and one output; output is blocked unless an input is energized (an OR gate), or unless all inputs are energized (an AND gate), or if no input is energized (NOT gate).

gateway

hardware and software that lets users on one network access resources on another network with a different communications protocol. Gateways operate at the 4th through 7th OSI layers.

gauge

a unit for measuring the diameter of the copper and aluminum wire in cables that connect customers to the central office. In measuring gauge, the lower the number, the thicker the wire. From the central office to approximately three miles along the cable, 26-gauge wire allows satisfactory transmission between the central office and the customers' telephone sets. Beyond three miles, successively thicker wire-24 gauge, 22 gauge, 19 gauge-- is required to supply good transmission.

GB

see GigaByte.

GBF/DIME file

also see Dual Independent Map Encoding.

Gbps

see Gigabits per second.

genlock

permits mixing and combining of two video signals.

genus

topological property that counts the number of holes in a surface: a doughnut has genus one--one cut is all that is required to remove the hole and deform the surface into a cylinder (which can then be developed to a part of a plane).

geocode

an index code based on spatial ordering of geographic entities--street addresses, subdivisions of homes, and so forth. In a GIS, a data value of a spatial object that provides information on its location and is used to access information concerning the object in the underlying database. One might use the tiles in Plate VI to develop a geocode.

geocoding
   translation of geographic coordinates into (x,y) grid cells.

geodesy
   science that seeks to determine by observation and measurement the precise shape, size, and dimension of substantial parts of the earth's surface and the earth as a whole. At a more local scale, it is concerned with precise locations of points on the surface of the earth.

geodesic
   curve along which to measure the shortest distance between two points; in the plane, this curve is a straight line. On a sphere it is a great circle (circle centered on the center of the sphere). Given two points, there may or may not be a unique geodesic joining them; in the plane geodesics are unique. On the sphere, two points that are not antipodal (at opposite ends of a diameter) have a unique geodesic joining them along the shorter of the two arcs into which they partition a great circle. If the points are antipodal, then there are an infinite number of geodesics (halves of great circles in this case) joining them.

   In any mapping situation, it can be quite important to know how distance is measured. In a GIS that calculates, behind the scenes, distances between pairs of points, it is often up to the user to ask how that distance is calculated--is it great circle distance or other distance? The answer can make a great deal of difference, especially when long distances are involved, in a planning context that might allocate funds as a function of distance. See also Manhattan geometry. In Plate VIII a Manhattan geodesic (there may be more than one) along the street grid is generally longer than the corresponding crow-flies geodesic suggested by the spiders (stars).

geodetic control
   network of   locations on the Earth's surface, surveyed and monumented according to established national accuracy standards.

geodetic elevation model of planetary relief
   model of planetary relief based on an octahedron (regular Platonic solid with eight equilateral triangular faces) or an icosahedron (regular Platonic solid with twenty equilateral triangular faces); the vertex sets of Platonic solids can all be embedded in the surface of a sphere and so can serve as natural tesellations of the sphere.

Geographic Base File/Dual Independent Map Encoding, GBF/DIME
   spatial data exchange format developed in the 1970s by the U.S. Bureau of the Census. See also Dual Independent Map Encoding.

geographic calibration
    aligning raster or vector data structure(s) with geographic coordinate system(s). The buffers in Plate X required aligning data structures with geographic coordinates in South Carolina.
geographic data
    data which have an embedded geographic location (such as latitude and longitude)
geographic entity
    occupies a position in space: attribute and locational data concerning the entity are recorded in the database of the GIS.
Geographic Information Retrieval and Analysis System, GIRAS
    USGS data files containing land use and land cover information for the U.S.A.
Geographic Information System, GIS
    a general term for an attribute data base management system linked to a topologically encoded land (and facilities) data base. Numerous types of GISs are available; see references for sources that list software with vendor names and addresses.
    One trend of the future seems to be that of integration: building GISs with on-board remote sensing capability, GPS equipment, and the like. Other directions for development, using theory to guide technology, also offer promise on the research horizon. See any number of the Plates.
geographic mean
    areally weighted mean.
geographic median
    areally weighted median.
geographics
    referring to coordinate systems, latitude/longitude or comparable geographic grid location references.
geophysics
    branch of physics that studies the set of physical forces that interact with the earth over time.
geoprocessing
    automated analysis of geographic data.
georeference system
    Cartesian coordinate system for locating points on the surface of the earth as a reference to locating points on a flat map.
geostationary satellite
    a satellite whose period of revolution in its orbit matches the period of time it takes for the earth to rotate on its axis; the satellite therefore has a view of the same part of the earth at all times. The radius of the

satellite orbit required to meet these conditions is about 30,000 miles. Also, geosynchronous satellite.

geosynchronous satellite
see geostationary satellite.

GFI
see Group Format Identifier.

GHz
see GigaHertz

giga
metric prefix for one billion (10 to the power of 9).

GigaByte, GB
computer unit: 10^9 bytes.

Gigabits per second, Gbps
a measure of transmission speed.

GigaHertz, GHz
a unit of frequency equal to one billion Hertz.

GIGO
see Garbage In, Garbage Out.

GIRAS
see Geographic Information Retrieval and Analysis System.

GIS
see Geographic Information System.

GKS
see Graphics Kernel System.

glitch [1]
an error caused by outside interference.

glitch [2]
a word used by programmers to explain away bugs.

global
a statement or symbol recognized throughout a computer process.

Global Positioning System, GPS
a system developed by the Department of Defense (U.S.A.), originally of nine satellites, but currently of 24 NavStar satellites (last one launched successfully, June 26, 1993) transmitting signals from space concerning precise positions on earth on a 24-hour a day basis. These signals often serve as the sources for inputs into GIS databases and have numerous other possibilities for application, including in vehicle tracking and intelligent vehicle highway systems, natural resource management, and urban planning. See specialized glossary in the next part of this handbook.

globe
spherical replica of the earth.

gnomonic projection
    see azimuthal projection.

GOS
    see Grade Of Service (transmission).

GPS
    see Global Positioning System.

Grade Of Service (transmission), **GOS**
    a measure of user satisfaction with the quality of transmission of a telephone conversation.

graded index fiber
    optical fiber with a variable refractive index; also called gradient-index fiber.

graded index profile
    profile showing continuous variation of the refractive index of a material (such as a glass fiber) from its core to its outer surface.

gradient
    actual slope (up or down); rate of change of a variable over distance. See Plate IX.

gradient analysis
    in a GIS, a procedure to determine the maximum rate of change of surface elevation in a Digital Terrain Model.

gradient filtering
    method of edge detection using two filters to obtain results superior to that of a single high-pass filter.

grain tolerance
    way to control distance between vertices.

graph theory
    graph theory is the most applicable branch of combinatorial mathematics today, with relevance as useful genuine mathematical models not only to virtually all areas of computer science, but also to the social sciences (anthropology, economics, geography, psychology, ...), the physical and life sciences (biology, chemistry, physics, ...), the engineering areas (civil, electrical, mechanical, ...), and the humanities (linguistics, literature, philosophy, ...). The reason for the pervasiveness of this phenomenon is that graphs capture precisely the structure of relationships between entities, i.e., networks of all kinds. The phrase, structural models, refers to those mathematical models which use graph theory. (Definition due to Frank Harary.) Also see connectivity. In Plate VII the bus routes are graphs: nodes with edges joining them.

graphic
    refers to a grouping of digital descriptions of map elements and relationships among those elements when represented as an image.

graphic overlay
   capability to superimpose one map on another, or multiples in data files, and display the result.
graphic tablet
   small digitizer.
graphic terminal
   workstation that can work with graphic files with only minimal local processing power.
Graphical User Interface, GUI
   a user interface that uses icons representing actual desktop objects that the user can access and manipulate with a pointing device; see WIMP.
Graphics Kernel System, GKS
   a graphic file and data interchange standard, designed to allow device independent programming (e.g., ANSI - GKS), cf. NAPLPS.
graphics software
   see illustration software.
graticule
   set of parallels and meridians on a map or chart.
grayscale
   image containing varying tone levels and intensities of gray; differences between adjacent tones represent differences between adjacent data levels. See Cover Plate.
greenness
   measure of vigor of biomass.
grid
   raster in GIS.
grid cell
   an area enclosed within the boundaries of a single unit of a lattice of intersecting sets of lines; usually refers to one in which spacing between the lines is uniform throughout and in which the angle of intersection of the two sets is a right angle--thus, the cell is a square. However, this need not be the case. See Plate VI.
grid format
   data structure in which data is stored in rectangular cells.
grid on a map
   two sets of intersecting lines superimposed on a map to create a coordinate system better suited to user needs than is the underlying graticule. The abstract tiles of Plate VI suggest a grid different from the coordinate system used to record data (presumably latitude and longitude).

ground control

locations on the ground for which there is accurate horizontal and vertical positional information.

ground range

horizontal distance from the projection of a flight path (ground track) to a given object.

ground resolution

in remotely sensed data, an image with ground resolution of 10 meters shows no feature on the earth's surface that is smaller than 10 by 10 meters.

ground track

orthogonal projection of flight path on the surface of the Earth; particularly of a satellite or other vehicle carrying remote sensing equipment.

ground truth

data on the earth, collected at the same time as remotely sensed data is being gathered about that site; serves to calibrate the remotely sensed data.

Group Format Identifier, GFI

header on a packet in X.25 packet-switched networks.

guardband

a gutter of bandwidth separating channels to prevent crosstalk between parallel send and receive channels.

GUI

see Graphical User Interface.

guy

a strong steel wire that looks like a metallic rope that is used to hold a pole in position. An arrow on the work print points in the direction in which the guy supports the pole.

H h

hachuring

any of various related ways of portraying land surface forms, using sequences of line segments, on paper maps; once in common use, now not that common. Family of Erwin Raisz still reprints Raisz's spectacular hand-drawn maps, many of which display complicated hachuring to portray land forms. In a hand-drawn block diagram, hachuring might be used to suggest the dark sides of the ridges; in the block diagram of Plate IX, the dark sides of the ridges are shaded assuming a light source in the upper left hand corner of the diagram.

Half-DupleX transmission, HDX

facility which permits transmission of information in either direction along network lines, but not in both directions at the same time--only in one direction at a time.

half-tap splice

used when a new piece of cable is spliced into an existing cable to reroute or replace an old section of cable and both cables remain in service for a short period of time.

halftone

continuous-tone image with gradations converted to discrete entities (lines or dots).

Hamming distance

a measure of distance between encoded sequences of 0s and 1s. The Hamming distance between two strings of equal length is the number of positions in which the strings differ. When a string is transmitted which can not be correct, one possible attempt to fix the transmission to replace it by a possible string whose Hamming distance from the transmitted string is as small as possible.

Techniques such as this are of practical importance in error detection and correction in the transmission of a signal over cable. They are of conceptual significance when used with other abstract structures. For example, a hypercube is a generalization of a cube to a dimension higher than 2--a situation that is difficult to visualize. A square has 4 vertices and 4 sides, each a line. A cube has 8 vertices and 6 faces, each a square. A 4-dimensional hypercube has 16 vertices and 8 hyperfaces, each a cube. In general, an n-dimensional hypercube has 2^n vertices; if each vertex is represented by a bit string of length n, there is a line between two vertices if the Hamming distance between their bit string representations is 1.

handhole

sub-surface chamber, too small for a man to enter, out where work may be done.

handshaking

preliminary part of a communications protocol that requires the interchange of predetermined signals between devices prior to making a connection.

hard copy

paper copy of online files--paper maps, text, and so forth.

hard disk

magnetic storage device for data and program files; see disk/disc.

hard drive

see hard disk.

hardware
those parts of a computer system consisting of the "hard," tangible, physical elements, such as mechanical, electrical, magnetic, or electronic components--auxiliary storage units, terminals, printers, and plotters--to name a few. Contrast with software and firmware.

hardware key
microcomputer device to authorize system configurations.

haze
an atmospheric condition which reduces the visibility of distant objects and which is caused by the scattering of the ultraviolet, violet, and blue waves of light.

HD
see High Density diskette.

HDDT
see High-Density Digital Tape.

HDLC
see High-level Data Link Control.

HDX
see Half-DupleX transmission.

head-end unit
hardware in a Local Area Network using split frequency bands to offer multiple services.

header
the first field in an entity record; control information at the beginning of a message.

heap
part of computer memory used for storing dynamic variables; greater RAM means greater heap.

hecto
metric prefix for 10 to the power 2.

heights along streams
GIS capability to interpolate points along streams at fixed heights. See Plate IX.

Hertz (Hz)
a term adopted as a standard unit for expressing frequencies (replaced "cycles per second"). This measure of bandwidth is equal to one cycle per second. Named for Heinrich Hertz.

heterojunction
a boundary surface in a laser diode at which a sudden change occurs in material composition across the boundary; there is often a level refractive index step at each such junction.

heuristic

simplification strategy that allows user to draw tentative, uncertain conclusions.

hexadecimal number

a number written in the base 16 number system; the first 10 numbers in the system are the integers 0 through 9; the last 6 "numbers" are the letters A, B, C, D, E, F.   Thus, any 8-bit byte can be characterized as two hexadecimal digits.

Hewlett Package Graphics Language, HPGL

a widely accepted format for exchanging graphics data between computers and output devices, especially plotters.

HFSP

see Human Frontiers Science Program.

hidden line removal

in GIS, removal of lines in perspective three dimensional drawing that should be obscured by the view angle or the solid itself. See Plate IX; the back edge of this solid block is not visible.

hierarchical

ordering scheme that moves from the global to the local, or from the general to the specific. Often, hierarchies can be represented as graph-theoretic trees with the trunk of the tree at the global end of the hierarchy; successive layers in the hierarchy see increased branching of the tree.

hierarchical database

database structure based on one-to-many (sometimes referred to as parent-to-child) relationships (from global to local) with pointers to define the relationships between record segments.

hierarchical switching

switching in a Local Area Network that is done in stages.

high boost filter

filter that improves image for a continuously varying raster object.

high level language

programming language, using statements close to natural language, used to develop software applications.   Also see fourth generation language. FORTRAN is an example of a high level language.

High Memory Area, HMA

in a DOS environment, the first 64 K of extended memory.

high pass filter

an operation that enhances high spatial frequencies, such as boundary edges, and attenuates (blocks) low spatial frequencies within an image; high pass filtering is used to bring out details difficult to see in the original.

High-Density Digital Tape, HDDT

analog tape system for data storage.

High-Density diskette, HD

storage medium; holds substantially more that a Double-Sided, Double Density diskette; also known as a Quad-Density diskette.

High-level Data Link Control, HDLC

an ISO T-1 rate communication protocol--the international standard communications protocol--compatible with CCITT X.25 level 2 recommendation.

HIS

see Hue, Intensity, and Saturation.

histogram

statistical tool; applied in a GIS context, shows frequency count of number of times a value occurs within a range of possible values.

HLS

see Hue, Luminance, and Saturation.

HMA

see High Memory Area.

Hollerith card

a keypunched, computer readable card containing 80 columns (now obsolete) adapted from Jacquard loom cards by Hollerith at IBM.

homeomorphism

topological mapping from one topological space to another; this type of mapping is structure preserving. It is useful in determining classes of topological invariants; the concept of genus is invariant under homeomorphism--a donut with one hole and a coffee cup with one hole in the handle are homeomorphic. Homeomorphic is the same as topologically equivalent. Concepts of this sort are important as stepping stones in constructing theory involving GIS.

homomorphism

algebraic mapping from one algebraic space to another; this type of mapping is neither one-to-one nor onto. It is quite general; see also isomorphism. Concepts of this sort are next steps in aligning GIS concepts with more abstract theoretical concepts.

horizontal accuracy

data position relative to control (i. e., source or input) in X and Y directions.

horizontal frequency

on a CRT, a measure in KHz of scanning frequency from one horizontal line to the next.

host

a central network computer in a data communications system that provides network services such as computation, data base access, and printer and file sharing access; mainframe and minicomputers are traditionally called hosts, servicing the needs of users who link to them via dumb terminals and PCs.

Hotline Oblique Mercator, HOM

map projection that is useful from spacecraft.

HPGL

see Hewlett Packard Graphics Language.

hue

attribute of color such as red, blue, or green, associated with a wavelength; these may be used to coordinatize a color system such as HIS or HLS. Also, any one of 100 pigments; first designation of the Munsell color system.

Hue, Intensity, and Saturation, HIS

system of characterizing color video output.

Hue, Luminance, and Saturation, HLS

system of characterizing color video output.

Human Frontiers Science Program, HFSP

a Japanese biological research team conducting research on artificial intelligence.

hydrography

science studying bodies of water and topography pertaining to drainage features.

hydrology

science studying distribution of water.

hypercube

hypercubes can be defined in many equivalent ways because they pervade both finite logic and contemporary parallel computers. In fact, in only a slight disguise, the finite boolean algebra with n atoms and hence 2^n elements is captured structurally by Q_n, the n-dimensional hypercube! The most intuitive description of Q_n is that of the unit cube in n-dimensional euclidean space, so that the unit interval is Q_1, the unit square is Q_2, the conventional box appearance displays the unit 3-cube, etc. (Definition due to Frank Harary.) Also see Hamming distance.

hypermedia

electronic capability to integrate graphics, text, audio, and other media.

hypsography; hypsometry

a branch of geography that deals with the measurement and mapping of the varying elevations of the earth's surface with reference to a datum, usually sea level; that part of topography dealing with relief or the devices by which elevations of terrain are indicated on maps; such as on Plate IX.

Hz

see Hertz.

I i

I/O

see Input/Output.

IC

see Integrated Circuit.

icons

picture symbols of elements in a graphical user interface; thus, a card game might be represented as an icon that is a picture of a deck of playing cards. When the icon is highlighted and opened, the entire game (word processor, graphics package, or whatever) appears.

ID

Idle Dead, see dead pairs.

IDIMS

raster image processing system developed for minicomputers by NASA.

idiographic class

set chosen with respect to specific characteristics of the underlying GIS data set.

IEEE

see Institute of Electrical and Electronic Engineers.

IGES

see Initial Graphic Exchange Specification/Standard.

ILD

see Injection Laser Diode.

illustration software

there is a vast array of illustration software for PCs and macintoshes producing (object oriented) images that print out on PostScript, and other, laser printers. New products abound; a trip to a computer software retail outlet, or a glance at a software magazine, will reveal a mere part of the diversity currently available.

image

encompasses all of photographs, television pictures, human vision and drawings, and more. The emphasis is on spatial linkages between underlying raster that convey meaningful information.

image analysis

process for locating boundaries in digital image data.

image map

an image that is made to look like a map; a satellite photograph may be made into an image map that will dovetail precisely with the corresponding USGS topographic map. The photograph in Plate III suggests a topographic map.

image processor

a circuit that converts an image, such as those produced from remotely sensed data, to digital form and then enhances it for further computer or visual analysis.

impedance

the total opposition a circuit, cable or component offers to alternating current.

import

process of sending data from a different system into the currently active system.

Improved Mobile Telephone Service, IMTS

pre-cellular analog mobile telephone service.

IMTS

see Improved Mobile Telephone Service.

in cable

under the serving area concept, the pairs in the feeder network are committed to a serving area and are terminated on the feeder (in) field side of the serving area connector.

in pairs

cable pairs terminated on connecting blocks and spliced into the feeder cable.

incident

in a graph, an edge joining two nodes is said to be incident with each of the nodes.

independent variable

see function.

index contour

a reference contour, always a specified multiple of contours, related to a given interval at a given scale.

index map

a reference map showing various boundaries, gridded by plat number--also shown are street names.

index-matching materials

light conducting materials used to reduce optical power losses. Refractive indices are chosen for interfaces that will reduce loss from reflection, dispersion, and so forth.

inductance

a property of a conductor or circuit which resists a change in current; symbolized by L.

inductor

a device which reduces transmissions loss in bridged station loops when one branch is idle. They are physically similar to loading coils, except for the type or core material.

information bit

a data bit.

Information Resources Manager (Management), IRM

deals with production, distribution, retrieval and other aspects of managing information.

infrared band

band of electromagnetic wavelengths between about 0.75 microns and the shortest microwaves of about 1000 microns. In remote sensing, this range is often partitioned into smaller subintervals of varying lengths.

infrastructure

the set of utility networks, such as water, sewer, electrical, and so forth that permit the distribution and removal of basic items required for effective human survival in communities. Often, it is displayed on a digital map such as the one in Plate V.

Initial Graphic Exchange Specifications, IGES

spatial data exchange format often used in CAD/CAM.

initialize

set program variables to their starting values (often zero).

Injection Laser Diode, ILD

a type of laser light source used to convert electrical signals to optical signals.

ink jet plotter

computer peripheral used to generate hard copy maps. Ink, of various colors, is forced through small jets into electrostatically-charged drops onto the paper in patterns guided by the computer.

input

data entered into a computer.

input device

hardware enabling input.

Input/Output, I/O

a measure of computer activity and data movement.

inset map

a small map at the margin of the large map; generally used as a supplement or modifier to the main map. Thus, a map of the state of Michigan might have a small inset map adjacent to the state boundaries showing the position of Michigan within the United States. Inset maps are thus often produced at a scale that is vastly different from that of the main map that they modify (as an adjective modifies a noun).

Instantaneous Field Of View, IFOV

Field Of View sensed remotely at an instant of time. Generally a narrow strip, its dimensions are described using angular measure related to the sensor or using ground area related to altitude of the sensor.

Institute of Electrical and Electronic Engineers, IEEE

professional and standards organization; responsible for standards 802.2 (data link layer) 802.3 (similar to Ethernet), 802.4 (Token Bus), and 802.5 (Token Ring) as part of IEEE Project 802, a family of Local Area Network standards. It issues its own standards and is a member of ANSI and ISO.

instruction

code that prescribes operation to be performed.

insulating joint

used in some buildings and manholes as protection against foreign voltage.

integer

whole number with only 0 after the decimal point; quicker to process than corresponding entries with non-zero values after the decimal point.

integrated circuit, IC

semiconductor containing several circuits.

integrated map

map of natural features in which individual map components are categorized under a variety of addresses--such a soil type and slope. See Plate II.

Integrated Services Digital Network, ISDN

a CCITT communications interface standard with a Primary Rate Access that allocates 23 bearer (B), circuit-switched channels of 64 kbps to voice, data, or video and 1 data (D) packet-switched channel of 64 kbps for control. Its focus is to integrate a wide range of voice and data issues over multiplexed communications channels. Provides end-to-end digital connectivity for simultaneous transmission of voice and data.

Intel

a chip (semi-conductor) manufacturer and one of the sponsors of Ethernet.

intelligence

a network or other electronic equipment with the capability, coming from built-in processing power, to allow it to execute complicated tasks using its firmware, is said to have "intelligence."

Intelligent Peripheral Interface, IPI
a type of high level computer-peripheral interface.

intelligent workstation
high-performance computer; often a 32-bit or 64-bit microcomputer.

Intelsat
INternational TELecommunications SATellite Organization.

intensity
refers to brightness of a color; second coordinate in HIS color definition scheme.

Inter-Record Gap, IRG
empty space separating data records on magnetic tape storage.

interactive graphics system
central computer and workstations from which a user drafts graphics in an interactive mode.

interactive processing mode
interaction in the processing is between user and processing unit; facilitates getting output quickly--as soon as the processing is complete.

intercell plot
a specified number of profiles or raster-files of each tying (adjacent) cell; all profiles or raster-files are perpendicular to the common profile or raster file and are plotted at the same scale as the map source.

interface [1]
a boundary shared between contiguous elements of an electronic system composed of various physical components; it is necessary, for example, to connect a printer to a minicomputer across a matching interface. The specification for communication between two systems is often across an interface; examples of interfaces are serial, parallel, SCSI, and GPIB.

interface [2]
any one of a large number of pieces of hardware used as a place to connect feeder cables to distribution cables with short cross-connection wires.

interference rejection
capability to remove transmission interference from a communications signal.

interlaced
refers to the refresh pattern on a CRT; in an interlaced pattern, odd rows of the CRT are refreshed and then the even ones (instead of refreshing rows consecutively). This causes difficulty with complicated images.

intermediate contour
a contour line between any two index contours.

internal data structure

organization of the reference linkages joining data elements in a system of data.

International Standards Organization, ISO

formed from government and industry representatives from various nations, whose OSI (Open System Interconnect) model for the structure of communications protocols is widely used. Membership is voluntary; ANSI is the U.S.A. representative.

International Telecommunication Union, ITU

telecommunications agency of the United Nations.

Internet

DARPA network used to share information and resources.

Internetwork Protocol eXchange, IPX

Novell's implementation of XNS protocols in Netware.

internetwork router

a device that functions in the OSI network layer for communications between subnetworks in a Local Area Network.

interpolate

to determine unknown values (such as elevations) between known values.

interpretation

identification and consolidation of classes of features in an image.

interpreter

a program (language processor) that translates and executes each high-level language statement before proceeding to the next statement; contrast with a compiler which translates the entire high-level language program into machine language before any statement is executed.

interpretive matrix

table of characteristics of mapped features used to present added information about the features.

interrogate

see query.

interrupt

a signal that tells the microcomputer CPU to suspend its current task in order to service a designated device or activity such as memory refresh.

Interrupt ReQuest lines, IRQ

lines used to send interrupt signals.

interval [1]

bounded set containing elements between endpoints; if both of the endpoints are included, the interval is closed; if both of the endpoints are not included, the interval is open; if one endpoint is included and the other is not, the interval is half-open (or equally, half-closed).

Knowing whether or not endpoints are included is important; some functions take on their maximum on the boundary, so it is important to know whether or not the boundary is included in the domain.

interval [2]

scale of measurement; distances between data points are fixed and based on an arbitrary starting point--such as the Fahrenheit temperature scale.

interval [3]

time separating two events.

inventory

collection of information concerning basic elements of a data set.

IPI

see Intelligent Peripheral Interface.

IPID

see Item of Plant IDentifier.

IPX

see Internetwork Protocol eXchange.

IRM

see Information Resources Manager (Management).

IRQ

see Interrupt ReQuest lines.

ISDN

see Integrated Services Digital Network.

island

in a GIS, a generalization of the common-sense notion of island--one unit of a particular land cover embedded within another larger one. In Plate II (lower right), there is a unit of purple land cover embedded within green land cover. Note also islands of blue within the red network in Plate VI.

island polygon

single individual polygon, such as an island or a lake.

ISO

see International Standards Organization.

isochrone

line connecting points of equal time distance from some given data point. The rate map in Plate XI can be used to draw isochrones. Also see rate map.

isochronous

data transmission scheme in which characters are separated by intervals of integral, rather than random (as in asynchronous transmission), length.

isoline
line connecting points of equal value--physical distance, time distance, barometric pressure, or any of a number of other types of values. The edges of the buffers in Plate X are isolines--all 10 miles from an Interstate.

isometric mapping
style of mapping that preserves distances between points--literally, iso=same, metric or distance.

isometry
geometric transformation in which one geometric structure is mapped to another in such a way that distances are preserved.

isomorphism
algebraic transformation mapping one algebraic structure to another via a function which is one-to-one, onto, and preserves the algebraic structure. Concepts of this sort are important in pointing out directions for theoretical research in GIS.

isopleth
see isoline.

item
a field in a feature attribute file.

item of plant
a single occurrence of a database entity; that is, a physical item of plant such as a telephone pole, cable, segment, or manhole; often shown on a digital map such as Plate V.

Item of Plant IDentifier, IPID
a unique number assigned to items of outside plant at the time of placement by a utility company.

iteration
repetition of process. Used to solve equations of certain sorts.

ITU
see International Telecommunications Union.

J j

jabber
nonsense transmissions (from corrupted data) from a station on a Local Area Network often causing the entire LAN to shut down.

jaggies
see line smoothing, aliasing.

Jet Navigation Charts, JNC
U. S. DMA 1:2 million scale map series standard--complete coverage of al the world's land areas. Used as a data source for DCW.

The ONC and JNC are products of a multinational project by the United States, Canada, Australia, and the United Kingdom.

JNC

see Jet Navigation Charts.

join

to connect two entities; sets or maps. On Plate II, three images are joined into one.

joint use

two or more utility companies (such as telephone, power CATV, and so forth) that enter a mutual agreement to use common poles, or trenches to provide service.

Jordan Curve Theorem

theorem from topology which states that a simple closed curve (does not cross itself) partitions the plane into three point sets: the inside of the curve, the boundary, and the outside. This theorem is critical in the digital mapping business, because it tells the user how to separate addresses on one side of the street from those on the opposite side.

To determine when points are inside or outside such a curve (a "Jordan" curve)--convex or concave--one can simply use the synthetic proof of the Jordan Curve Theorem.

Step 1. Preselect a direction and mark it in the plane with an arrow (a north arrow, for example); this orientation should be such that it is not parallel to any sides of the given polygon.

Step 2. Draw rays from each of the given points, oriented in the preselected direction.

Step 3. Assign a parity to each point--parity is the number of times the ray crosses a polygon edge. If a ray hits the polygon at a vertex of the polygon, it adds to parity if and only if the edges forming the vertex lie on opposite sides of the ray.

Step 4. Points with even parity lie outside the polygon; points with odd parity lie within the polygon (zero is even).

Because the Jordan Curve Theorem is fundamental to the theoretical structure underlying digital mapping, it seems likely to be fertile ground as a conceptual base from which to explore other abstract research directions; one that might be fruitful, since this proof relies on the notion of parallelism, is to consider possible analogues in non-Euclidean electronic (or other) spaces.

joystick

hand-controlled lever for cursor movement on a graphics CRT.

jumper

connector on a microcomputer circuit board joining two adjacent terminal posts; a wire used to connect an "in" pair to an "out" pair in a terminal.

junction box
aerial cross box.
justification
refers to position of text or symbol; the text in the terms and concepts on this page is right-justified--it extends all the way to the right-hand margin.

K k

K-means
iterative clustering method; quantitative tool in which an n-dimensional raster is partitioned into K clusters.
kappa
Greek letter conventionally used to denote elevation as the third component in an n-tuple describing geographic location.
Kauth's Tasseled Cap
quantitative method based on greenness, brightness, and wetness of a site to reduce, using linear combinations of satellite spectral bands, the satellite images to maps of the site.
KB
KiloByte (10^3 bytes).
Kbps
see Kilobits Per Second.
kernel
entire content of command-interpreting templates is a kernel; this is like the algebraic notion of kernel in which the kernel of an algebraic transformation consists of those elements which map to the identity element of the algebraic structure; here, one always can come back to the kernel (or shell).
key
item used to identify other items, as in an indexing key. See relational data base.
key entry
technique employed to enter nongraphic data into a computer.
key file
file containing information or codes about means to access data.
keyboard
peripheral for entering input into a computer; many keyboards are QWERTY keyboards--the same basic pattern of alphabetic letters as on a typewriter keyboard. See also QWERTY.
Keyboard Send/Receive, KSR
teleprinter transmitter and receiver with transmission capability from the keyboard only.

KHz
   see KiloHertz.
kilo
   metric prefix for one thousand (10 to the power of 3).
Kilobits Per Second, Kbps
   a measure of transmission speed.
KiloHertz, KHz
   thousands of Hertz.
knowledge base
   component of an expert system that contains facts and heuristics.
Kriging
   interpolation to obtain estimates, that are statistically unbiased, of surface elevations from a set of control points (named for D. G. Krige).
KSR
   see Keyboard Send/Receive.

L l

label
   in a GIS, a vector element containing text that identifies a node, line, or polygon. In a raster GIS, user selects names for feature types to use as labels for particular areas of the image. See Plate V.
lag
   time or space interval reflecting some sort of delay.
lambda
   Greek letter conventionally used to denote longitude as the second component in an n-tuple describing geographic location.
LAN
   see Local Area Network.
land cover
   natural materials, such as water or vegetation, that are present in a region. See Plate II, lower right side.
Land Information System, LIS
   GIS with a data base focused on land records describing physical and legal characteristics of areas.
land use
   type of human function which determines how a parcel of land is used: urban, agricultural, and so forth. See Plate II, upper right side.
Landsat
   a program of earth-observing satellites initiated by NASA in the early 1970s (first satellite launching in 1972); since 1986, Landsat has been owned by Earth Observation Satellite Company (EOSAT).

Landsat satellites relay much information that is useful in the digital mapping environment.

LAP

see Line Access Procedure; see Line Access Procedure, Balanced.

LAPB

see Line Access Procedure, Balanced.

LAPD

see Link Access Procedure for the D-channel.

LAPM

see Link Access Procedure for Modems.

laptop

portable computer weighing less than six pounds.

large scale

a map of scale with a relatively large representative fraction; 1:5000 is larger than 1:10,000. In the corresponding maps of the same physical dimensions, the area covered in the large scale map is not as extensive as in the smaller scale map; however, the coverage in the larger scale map can be made to show much more detail.

Large Scale Integration, LSI

refers to integrated circuits with 100 to 5000 logic gates or 1000 to 10,000 memory bits in one silicon chip; apart from the cost reduction by elimination of separate components there are advantages in the reduction of size and power consumption and in increased reliability. Circuit ranges are: Small (SSI), 2-10 circuits; Medium (MSI), 10-100 circuits; Large (LSI) 100-1000 circuits; Very large (VLSI), 1000-10000 circuits; Ultra large (ULSI), over 10000 circuits.

LASER

see Light Amplification by Stimulated Emission of Radiation.

laser diode

a junction diode.

laser plotter

a plotter that uses a laser to expose photosensitive film.

laser printer

a computer peripheral that uses a laser to apply charges to a rotating drum which then picks up toner at locations guided by computer input.

lashing

wire wrapping that attaches a cable to a strand.

lasing

an action that takes place in a laser of which the eventual result is the release of lightwaves.

LATA

see Local Access and Transport Area.

lateral cable

a cable that comes out from a manhole (often in a conduit) and goes to buried or aerial cable.

lateral duct

a duct that goes from a manhole to a structure other than a manhole or central office.

latitude/longitude

global coordinate system; locations are given as geographical coordinates that represent angular measurements relative to some surface approximating the surface of the earth (sphere or ellipsoid). Latitude measures the angular degree distance along a meridian of a point on the Earth north or south of the Equator. When a sphere is used as the model for the Earth, 1 degree of latitude translates to about 69 miles on the surface of the earth. Longitude is the angular degree distance measured along the Equator or a small circle parallel to the Equator, of a point on the Earth east of west of the Prime Meridian passing through Greenwich, England. On the Equator, 1 degree of longitude translates to about 69 miles on the Earth-sphere. Parallels are circles each point of which has the same latitude; meridians are halves of circles each point of which has the same longitude. Latitude is conventionally denoted by the Greek letter phi; longitude by lambda. See maps in Figures 1-24.

lattice

a framework of crossed lines (generally; also has a more technical mathematical interpretation); often called a fish net, in a GIS context, when draped over a three dimensional surface. See Plate I.

law of the excluded middle

basic logical law, which states that a statement is either true or false; it may not hold any sort of partial truth value between these two extremes. Most of modern mathematics and computer science is based fundamentally on this law; binary states are reminiscent of this law; thus, it may be a useful direction to consider for theoretical research in digital mapping.

layer

digital map equivalent of a manual map overlay; also known as coverage (ESRI), theme (ERDAS), level (Intergraph), channel (Scitex), *et cetera*. Each database layer, containing homogeneous map features, is registered positionally to other layers so that good superimposition is the result. Plate VII has two clear layers: one of bus routes and one of a street grid.

In a telecommunications environment, there are various strategies for layering communications protocols. In the OSI model, Layer 1 is the physical layer, including physical signaling and interfaces; Layer 2 is the data link layer for local addressing and error detection; Layer 3 is the Network layer for handling internetwork addressing and routing; Layer 4 is the Transport layer for error correction to guarantee error-free delivery; Layer 5 is the Session layer for connections to application programs; Layer 6 is the Presentation layer for format and code conversion; and, Layer 7 is the Application layer offering an interface with network users.

IBM's Systems Network Architecture uses layers to describe data communications systems and facilities in a standard manner. The architecture is partitioned into an application layer, a functional management layer, and a transmission subsystem layer and is designed to allow considerable flexibility in which end users are independent of specific data communications system services used for information exchange.

Also, used as in level of power in a signal.

layout

the manual procedure in which items of plant are transferred from existing manual PLRs to the mechanized land base.

layover

on a radar image, the displacement of the top of a tall feature with respect to its base.

LCD

see Liquid Crystal Display.

leased line

telephone line for the exclusive use of the customer who rents it.

least squares

often-used quantitative method to fit a line or curve to a finite set of points. It minimizes the sum of the squares of the errors for each element of the data set; distance from data points is measured along a vertical line. Because it is distance squared that is minimized, outliers can play a significant role in skewing the results.

LED

see Light Emitting Diode.

LEF

see Local Event Flag.

leg of cable

a portion of cable from a splice to either the next branch splice or the end of the cable.

legend

map component explaining meaning of symbols used on the map.

LeTteRs Shift, LTRS

Baudot Code shift that permits the printing of alphabetic characters.

level

see layer.

LF

see Line Feed.

library

in a computing context, a set of standard subroutines or symbols in digital form. Plate VI displays a library of coverages for a river system.

Light Amplification by Stimulated Emission of Radiation, LASER

a coherent-light generator in which molecules of certain substances can absorb incident electromagnetic energy at specific frequencies, store the energy for short periods in higher energy-band levels and then release the energy, upon their return to the lower energy levels, in the form of light at particular frequencies in extremely narrow frequency bands; the release of energy can be controlled in time and direction to generate an intense, highly directional narrow beam of electromagnetic energy that is coherent.

Light Emitting Diode, LED

a device used to convert electrical signals to optical signals. More reliable than an incandescent lamp.

light pen

hand-held photosensitive interactive peripheral; used to manipulate images or text on a refreshed computer screen--a pen-sized cursor used to point on a computer screen.

lightwave communications

branch of telecommunications focusing on the development and use of equipment using visible, or near-visible, electromagnetic waves for communication.

limit line

adjustable graphic line separating values that are used from those that are not.

line

geometrically, a one-dimensional object. In GIS terminology, a segment joins two points; a string is formed from a connected sequence of line segments; an arc is the set of points derived from a well-defined mathematical function; and a chain is a directed sequence of nonintersecting line segments or arcs with nodes at each end. Line segments intersect if and only if they have a point in common, in the geometry of the GIS. Thus, a green line in one spider (star) in Plate VIII intersects the underlying street grid at most at two nodes--at either end of a green edge.

Also, a physical communications paths in a computer network or telecommunications network.

Line Access Procedure, Balanced, LAPB
    an error correction protocol.
line artwork
    art composed of discrete entities as opposed to continuous tone art.
Plate XI conveys its image using discretely colored entities.
line driver
    in telecommunications, a device to convert a signal to ensure its
reliable transmission beyond initial signal limits.
line editor
    a program that allows you to make additions and deletions to a file
on a line by line basis.
Line Feed, LF
    control character (ASCII or EBCDIC) to move to the next line.
line follower
    peripheral using a laser to trace lines from a source map and
transform them into digital form.
line in polygon
    to determine which polygons in a set contain segments from a layer
of superimposed segments. See the entry for line, and the discussion
involving Plate VIII.
line length calculation
    computational procedure to calculate distances between successive
coordinate pairs linked geometrically.
line mapping
    mapping by (x,y) coordinates, joining them with line segments--used
to depict linear features--such as road networks; see Plate XI.
line of sight
    an uninterrupted line between two points, each of which is visible
(along the line) from the other.
line on polygon
    an overlay of a set of segments on a set of polygons. The lines are
truncated at intersections with polygon boundaries. See Plate V.
line printers
    computer peripheral that prints text and images on a line-by-line
basis; a dot matrix printer is an example of a line printer.
line smoothing
    automatic procedure which creates a new set of points to define a
line (more smoothly), according to user-defined tolerance.   Also,
jaggies, anti-aliasing.
line string
    a line element in a digital graphics file.
line thinning
    reduction of the number of points defining a line.

line turnaround
   reversal of transmission direction on a half-duplex telecommunications circuit.
line weight
   the width of a plotted or drawn line.
linear array
   type of scanning sensor in which a line of data is obtained from an array of detectors.
linear programming
   mathematical theory which uses the fact that when the boundaries of a region are linear, a linear function defined over that region will assume its maximum (or minimum) at a vertex of the region. For example, if the function $f(x,y) = 2x + 3y$ is defined for the interior and boundary of the triangle with vertices (1,1), (3,2), and (5,0), its maximum value over the entire region is 12 (at (3,2)). Normally, the boundaries are defined by linear inequalities, and the vertices must be determined. In systems with many variables, there are often too many vertices to consider evaluating the function at all of them. A method known as the simplex method simplifies finding the correct vertex by manipulation of a matrix. Computer algorithms exist to do this manipulation in very large problems.
linear transformation
   algebraic transformation that maps lines into lines; linearity is preserved under this transformation. Theorems about linear transformation, and the theoretical structure that they form, are the subject of the algebra of vector spaces: linear algebra.
   In a GIS context, the idea of a linear transformation is applied to align overlays of maps: map registration.
linearity
   in a telecommunications environment, the property of equipment that allows it to carry a signal without introducing distortion.
link
   to join; as in, to join separate modules into an executable program.
Link Access Procedure for the D-channel, LAPD
   an ISDN layer 2 protocol (Q.921).
Link Access Procedure for Modems, LAPM
   an error correction protocol.
linker
   a program that creates an executable program, called an image, from one or more object modules produced by a language compiler or assembler; programs must be linked before they can be executed.
Liquid Crystal Display, LCD
   a computer screen display technology (cf. EL).

load of coil
an electronic device added to a cable pair or pairs that assists in bringing about a good, clear hearing telephone line.   Improves transmission over long distances.

load point
location of load coil; occurs at 2500' and then subsequently at 3000' intervals.

loaded cable
cable with uniformly spaced load coils to improve transmission.

loaded line
telephone line containing loading coils that minimize voice-frequency amplitude distortion.

local
a modifier that indicates a small geographic area, as contrasted with global.

Local Access and Transport Area, LATA
geographical subdivisions of the U.S.A. to define local telephone service--as opposed to area codes which define geographic regions of long distance service.

Local Area Network, LAN
a computer connection system providing for shared access to and movement of data over some limited geographic area such as a building complex or a university campus.   Typically, a LAN is owned by its user, although it may have gateways to various public and private networks.

Local Event Flag, LEF
a computer operating system message.

local loop
communications channel, such as a telephone line, linking a subscriber to the central office.

local printer
as opposed to a network printer, a printer attached directly to a port of a computer (for example).

location theory
that branch of geography that employs mathematical (or other) theory to make decisions about, and to analyze patterns of, real-world location: Thünen's concentric ring model, Hotelling's beach vendor problem, and Christaller's central place theory are classical approaches to location theory that are still fruitful.

A class of problems, referred to as location/allocation problems in which the object is to find optimal (or good) sites from which to distribute a service (to which to attract clientele), are also part of location theory. See Plates IV, VII, VIII and the more abstract Cover Plate.

lock

a software mechanism that limits access to a specific microcomputer resource (e.g., expansion board or microprocessor).

locking

method of protecting the integrity of shared data on a LAN; a multiuser database uses record-locking to prevent more than one user from concurrently using the same record.

log file

a file containing the history of the work.

log-on, log-off

to establish/terminate communications between a terminal and computer or within a computer with access restrictions (same as log-in, log-out).

logic

the study of reasonable presentation of fact in a consistent manner; the base of communications. One often speaks of inductive and deductive logic; deductive logic is the base of mathematics which is the root of computer science. Statements in deductive logic can be expressed in terms of a set of fundamental variables and operators: and, or, not; and quantifiers: for all; there exists. A number of other abstract concepts are useful--to name a few--if..then; if and only if (iff), converse, inverse, and contrapositive. Truth tables, which offer a tabular structure for the data of logic, are a useful way to systematically establish the validity of statements in deductive logic.

logical channel number

in a packet-switched network, a number assigned when a call is placed.

logical contouring

the interpolation of contours by logically spacing the distance between existing contours and/or making reference to other known features such as benchmarks, lakes, depressions, spot elevations, stream and river patterns.

logical links

in a DEC environment, a link carrying a stream of data between two user-level processes.

logical name

a character string used to refer indirectly to files or devices.

Logical Unit 6.2, LU6.2

Logical Units are IBM's network-defined paths to computers or devices like printers; LU 6.2 is the Logical Unit type that implements

IBM's APPC peer-to-peer (or device-to-device) networking scheme under SNA.

login

see log-on.

longitude

see latitude.

Longitudinal Redundancy Check, LRC

an error detection method; also called horizontal parity check.

look-up table

array to convert data from one form to another, designed as a convenience for the GIS user.

loop

the connection between a subscriber's telephone and the central office. It takes its name from the pair of wires that constitutes a loop of direct current from the central office to the subscriber for signaling.

loop carrier

local carrier system using multiplexed digital circuits to connect a central office to remote terminals (often to over 100 subscriber lines).

loop makeup

a procedure in which a count is traced from the central office to a location keyed in by the user. The output is the distance from the central office, and the gauge and length of the cables through which count traverses.

loopback

diagnostic test that returns the transmitted signal to the sending device after it passes through a set of stations on the network.

low dimensional topology

the direct theoretical basis of GIS: there are individuals at a number of institutions who are interested in this. Some of the early ones that captured our attention were at the University of Maryland; Long Island University; and Carnegie-Mellon University.

low pass filter

an operation that enhances low spatial frequencies and attenuates (blocks) high frequencies; as a result, detail is smoothed and reduced within an image.

low-level programming language

programming language that works at a level that is close to the computer's actual instruction set to work with the basic electronic components of the computer.

LRC

see Longitudinal Redundancy Check.

LRS

Land Records System. See Land Information System.

LSI
   see Large Scale Integration.
LTRS
   see LeTteRs Shift.
LU 6.2
   see Logical Unit 6.2.
luminosity
   lightness or darkness of a color.

M m

"M" Field Code
   maintenance (changes, rearrangements).   Replace old type with newer model. Also called "M" operation--telephone context.
M1-2
   a multiplexer that combines 4 DS-1 signals into a DS-2 signal.
MAC
   see Media Access Control.
machine language
   an operational binary code that a computer can recognize and to which it can respond without further translation.
macro
   a group of instructions that the programmer defines and names once, at the beginning of a program, and then invokes the execution of many times during the program simply by referencing the macro's name. It represents an application of a one-to-many transformation: one command represents an entire sequence of instructions.
macro programming
   process of creating macros; see macro.
macrocell
   geographic territory, larger than a microcell, that functions as a transmitting and receiving area within a delivery architecture for a cellular phone system.
magnetic tape
   medium on which digital data can be stored sequentially.
magnetometer
   instrument to measure magnetic fields.
Main Distribution Frame, MDF
   CO switch wiring location.
main frame
   in telephone terminology, a location where all cables are terminated within a central office.

main frame verticals
cable riser in central office; terminals outside plant.
main memory
see core memory.
mainframe
a large computer complete with peripherals and software manufactured by a single vendor. Mainframes often have a closed architecture, and they support many users.
Manchester encoding
binary signaling procedure combining data and clock pulses.
Manhattan geometry
the geometry of a rectangular grid--as found on a CRT with square pixels, graph paper, a digitizer, or a city street network. The fundamental difference between Manhattan geometry and plane geometry is that geodesics in Manhattan geometry are not unique whereas in plane geometry they are. See also geodesic. See Plate VII; the particular street pattern here is part Manhattan, part not--a typical situation.
manhole
an underground vault which is large enough for a person to work in, and into which cables enter through conduits. Manholes divide conduit runs into sections permitting easy cable installation and splicing; they might be displayed on a digital map such as Plate V.
manifold
device permitting access to pressure pipes.
Manufacturing Automation Protocol, MAP
a subset of OSI network and communications protocols designed for computer-assisted manufacturing. A token-passing bus deigned by General Motors Corporation for factory environments. IEEE standard 802.4 is similar to MAP.
manuscript
an original source drawing map as compiled or constructed from various data.
map
representation of the earth (or other planet, for example), using a replicable mathematical procedure, on a surface (often a flat one). More specifically, a graphic and/or coordinate representation, usually on a plane surface and at an established datum and scale, of natural and artificial features on the surface, and their spatial relations, of a part or the whole earth or other planetary body. See section of Plates.
MAP [1]
see, Manufacturing Automation Protocol.

MAP [2] - map scale 165

MAP [2]

see Map Analysis Package.

map accuracy (absolute)

deviation between a true location on the earth and the scaled location on a map.

map accuracy (relative)

distance deviation between two points on the earth and that distance as scaled on a map.

Map Analysis Package, MAP

program by C. D. Tomlin to analyze spatial data coded in grid cells.

map generalization

reduction of detail on a map in order to present a clearer picture; this process involves the human characteristic of judgment and is at present difficult for a computer to execute well.

map grid

see grid on a map.

map projection

mathematical transformation of the earth's surface into the plane; the coverage can never be perfect (one cannot flatten out the peel of an orange without ripping it). Thus, compromises must be made and hence it becomes important to distinguish which projections have which characteristics. A projection that does not preserve area would be misleading if a global agency were to use a world map to compare forested land in Brazil and Canada--when the area in Brazil is accurately represented but that in Canada is badly distorted to appear twice as large. Funding allocations for reforestation might also represent the distortion. Thus, choose a map projection with characteristics suited to fit the problem at hand.

See particular projection classes, such as azimuthal projections, conic projections, cylindrical projections, and others. Also, note maps displayed in conjunction with selected projections; see, Albers, azimuthal, Bonne, Behrmann, conformal, conic, cylindrical, equal area, Mercator, and others.

map registration

see linear transformation.

map scale

ratio of units of linear measure on the map to units of measure on the earth. Ways of representing the map scale ratio:

representative fraction

one unit on the map represents 50,000 units on the earth is written as a representative fraction as 1:50,000. Thus, 1 inch on the map represents 50,000 inches on the earth, or 2 centimeters on the map represent 100,000 centimeters on the earth. It is very useful to have this method of measurement that is independent of units; however, when one enlarges or reduces a map, one must be certain to alter the fraction accordingly, as well.

bar scale

scale can also be expressed in units; thus, one inch might represent 100 miles. This can simply be stated in words, or it can be expressed visually by drawing a line segment one inch long and labelling the left hand endpoint with a zero and the right hand end point with 100--then, insert the word "miles" over the bar. This sort of scale requires alteration when one wishes to switch from miles to kilometers, for example, but it does not require alteration when zooming in or out; the enlarging or reducing procedure also alters the scale correctly.

MAP/TOP

see MAP and TOP.

MapInfo

GIS from MapInfo; see references in front matter.

marginalia

see digital marginalia.

mark

represents a binary numeral 1.

mask

see data mask.

mass storage device

a computer peripheral to increase storage capacity, such as a tape drive. Also called auxiliary storage.

master station

a monitoring station in any of a variety of circuitry environments: LANs, multipoint circuits, point-to-point circuits.

material dispersion

in optical fibers, the variation in the refractive index of a transmission medium as a function of wavelength.

math coprocessor

any computing device designed to help a general-purpose computer perform math operations.

matrix

a rectangular pattern (grid) expressing elevation posts (numerical model) at a given spacing.

matrix printer

printer that forms output patterns as a sequence of dots from (for example) a 10 by 10 grid of available dots for each character; also known as dot matrix printer.

maximum acceptance angle

see, acceptance angle. In an optical fiber, the sine of the maximum acceptance angle is the square root of the difference of the squares of the refractive indices of the core glass and the cladding of the fiber.

maximum likelihood classification

image classification scheme involving multivariate images; if there is likelihood of confusion between which class an image belongs to, then it is assigned to the class of highest probability.

MB

also see MegaByte (roughly 10^6 bytes).

Mbps

Megabits per second, Millions of bits-per-second; see Megabit.

MCVD

see Modified Chemical Vapor Deposition Process.

MDF

see Main Distribution Frame.

mean

one measure of central tendency; an average.

Mean Time Between Failures, MTBF

a statistical measure of the expected reliability and integrity of a computer system or individual components--indicates average duration of periods of time with fault-free operation.

Mean Time To Repair, MTTR

a statistical measure of computer integrity used in conjunction with the Mean Time Between Failures to derive availability figures.

measure

ability to determine the size (possibly the shape) of geometric objects. A number of objects can be measured in a GIS context: usually, angle, linear distance, some curvilinear distance, perimeter, areas, slopes of lines, numbers of items, and volumes. There are various conventional terms associated with types of measurement scales: interval, nominal, ordinal, and ratio.

Media Access Control, MAC

controls LAN traffic to help avoid data collisions as packets move on and off the network through the adapter card; a subset of the OSI data link layer.

medium

physical substance on which data can be stored; magnetic tape, map, and so forth.

meet-me bridge

a type of telephone bridge that can be accessed directly by calling a certain access number; it provides dial-in teleconferencing; the term "meet-me bridging" refers to the use of this type of bridge. Also, meet-me teleconferencing.

mega

metric prefix for approximately one million (2^{20}).

Megabit

a measure of transmission speed.

MegaByte, MB

a measure of computer memory equal to 1,048,576 bytes = 2^{20} bytes.

megaflops

see MFLOPS.

MegaHertz, MHz

a unit of frequency equal to approximately one million Hertz.

member function

in object-oriented programming, a procedure that implements a message.

memory

locations in which computer stores, and can retrieve, data and sets of instructions.

memory resident software

program that resides in memory and allows other programs to run normally. Also Terminate and Stay Resident, TSR.

menu

a display of command options on a computer screen; the user selects from the display by pointing to the options using a cursor, or similar tool, but without typing.

Also, a graphic and text display, linked electronically to a digitizing station, which represents the operations a digitizing technician may perform on the contents of a design file.

In a telephone context, a menu provides a means for the graphics work station operator to interface with IGDS.

Mercator projection

projection centered along the equator with evenly spaced meridians. Meridians and parallels are perpendicular; lines of constant compass bearing are preserved so that the Mercator projection is valuable for conventional navigation of a ship. It is not an equal area projection; landmasses poleward are greatly exaggerated. Thus, it would not serve as a reasonable map on which to compare visually extent of forested lands in northern Canada and Brazil. Concerns of this sort are important to note when making policy decisions concerning allocations of funding for issues involving forested lands. See cylindrical projection.

merge

to join items from two or more distinct sets whose internal ordering structure is sufficiently similar to perform the splicing.   Plate II illustrates splicing of maps.

meridian

a half of a great circle passing through the Earth's terrestrial poles. The prime meridian is the reference line from which others are measured using longitude. The 180th meridian is the International Date Line.   Meridians are measured in degrees east or west (0 to 180 degrees) of the Prime Meridian. See Figures 1-24.

message

a complete transmission of information.   In object-oriented programming, a command request for an object to be executed on the data.

Metal Oxide Semiconductor, MOS

MOS devices of all types depend on the conditions at the surface between a semiconductor layer - usually silicon - and an insulator layer.

metes and bounds

method of describing the boundary (bounds of land by a sequence of lines identified by a measure of length (metes) and direction from the previous line.

method of least squares

see least squares.

metric prefixes

prefixes in the metric system of measurement which indicate multipliers of powers of 10 with positive or negative exponents.

MFJ

see Modified Final Judgment.

MFLOPS

see Million FLOating Point operations per Second.

MFM

see Modified Frequency Modulation.

MHz

see MegaHertz.

micro

metric prefix for one millionth (10 to the power of -6).

microbending loss

loss of power in an optical fiber due to bends or kinks in the the direction of the fibers.

microcell

relatively small geographic territory that functions as a transmitting and receiving area within a delivery architecture for a cellular phone system.

microcellular

small cellular phone system serving users within some pre-defined radius (e.g. 1000 feet) of the base station which transmits and receives calls.

microcomputer

a microprocessing system; often a desk-top or lap-top computer.

micron

abbreviated by the Greek letter mu; a micrometer.

microprocessor

a computer on a silicon chip.

microwave

short electromagnetic wave between 1 meter and 1 millimeter in wavelength. Used in microwave relay systems and radar.

MIL-Spec

MILitary specification of standards for vector map products.

milli

metric prefix for one thousandth (10 to the power of -3).

Million FLOating Point operations per Second, MFLOPS

a measure of computer processing speed (also known as megaflops).

Million Instructions Per Second, MIPS

a measure of computer processing speed (or "Misleading Indicator of Processor Speed").

minicomputer

a medium-scale computer, often operated with interactive dumb terminals. As contrasted with a mainframe on the large end and a microcomputer on the small end.

minute

one sixtieth of one larger unit: of one hour, of one degree of latitude, or of one degree of longitude. U.S. Geological Survey quadrangles are commonly either 15 minutes or 7.5 minutes. Minutes are denoted using a quotation mark; thus, 30 minutes is often written as 30".

MIPS

see Million Instructions Per Second.

modal dispersion

changes in the relative magnitude of the frequency components of a wave during the propagation of an electromagnetic wave.

modal loss

loss of energy in an open waveguide. The cause for the loss may be due to a variety of factors.

mode

in a communication system, a medium for the transmission of digital and other signals. In an optical fiber, an arrangement of electromagnetic waves in a transmission medium. The number of modes that a given fiber can support is a function of fiber radius, the refractive indices of the core and cladding of the fiber, and the wavelength.

mode conversion

switching from one mode of propagation of electromagnetic waves to another.

model

use of a manageable representation of a particular problem, be that representation a tangible one, such as a globe, or an abstract one such as a computer model. Plate IX shows a three-dimensional model; Plate VIII shows a graphical, structural model of linkage (in the spiders/stars).

MODEM

see MOdulator-DEModulator.

modem eliminator (emulator)

a device to connect a local terminal to a computer port, instead of the expected pair of modems. Various cable and connector arrangements can execute this function.

Modified Chemical Vapor Deposition process, MCVD

process for the production of optical fibers; contrasts with OVPO process; same as modified inside vapor phase oxidation process.

Modified Final Judgment, MFJ

the primary legal instrument defining contemporary deregulation in the telecommunications industry.

Modified Frequency Modulation, MFM

a disk drive data encoding technique.

modular arithmetic

in packet-switched networks this form of arithmetic is used to count the maximum number of states for a counter and to describe various network parameters. An arithmetic system that functions modulo 8 is one in which the count proceeds from 0 to 7 and then when 7 is exceeded is reset. Thus, 7+2(modulo 8) is 1(modulo 8). Generally, two integers a and b are said to be congruent relative to a modulus m if and only if there exists a constant k such that a=b+kn; in that case, a is said to be congruent to b mod n. In the example above, 9 = 1+1*8, so that 9 is congruent to 1 mod 8. The modulus of 8 appears in 8 bit processors and hexadecimal code.

modular splicing

method of splicing 25 pairs at a time in one module.

modulation

process by which a carrier is modified in order to represent an information-carrying signal.

MOdulator-DEModulator, MODEM
a device that is a modulator in one direction and a demodulator in the other; in frequency division carrier systems, modems change the frequency with voice band on one side and carrier frequency with either erect or inverted sidebands on the other side. It is also called a data set.

module
a subdivision of a program that performs one or more software functions and can be connected with others to form a system. (May also be used in a similar context in conjunction with hardware.)

modulo
see modular arithmetic.

MOESI
the five cache state attributes of the Futurebus: M = Modified; O = Owned; E = Exclusive; S = Shared; I = Invalid. See bus.

moiré
pattern of light and dark which occurs in certain orientations of superimposing one screen on another; often wave-like patterns; sometimes jarring to the eye. Some screen savers use these patterns; silk fabric, when draped in various planes often displays these otherwise unsuspected patterns.

monitor
a display device; often a Cathode Ray Tube or LCD.

monitor station
on a Local Area Network with a ring topology, the station that checks to see that the ring stays intact.

monochrome image
a single color display of an image.

monument
permanent marker on the site of a survey point.

mortality
the year in which equipment is placed (utility company, cf. in-service data).

MOS
see Metal Oxide Semiconductor.

mosaic
large image assembled from small pieces (tiles)--in a GIS environment, from raster objects. Also, assembly of aerial photographs or maps best fitted together. Also see, tiling.

motherboard
the main printed circuit board which resides within the computer; this main board usually contains the CPU and areas, if required, for additional printed circuit cards, boards or modules to be connected.

mouse
a hand-held pointing device that controls the movement of a cursor on a screen of a graphics monitor. Buttons on the mouse are often used to edit and enter graphics.

moving average
sequence of averages that can be taken in various ways.

MSA
Metropolitan Statistical Area (obsolete) (see SMA)

MSI
Medium Scale Integration; see Large Scale Integration

MTBF
see Mean Time Between Failures.

MTTR
see Mean Time To Repair.

MULDEX
see MULtiplex/DemultiplEX.

multicast bit
in Local Area Networks, a bit in the Ethernet addressing structure used to indicate that a message is to be broadcast to all network stations.

multimedia
combination of sound, video, graphics and various media.

multimode fiber
optical fiber waveguide in which more than one mode can propagate.

multimode group-delay spread
the variation in group delay in an optical waveguide among the propagating modes supported on a single frequency.

multimode laser
laser that emits radiation at a number (greater than or equal to two) of wavelengths.

multimode waveguide
a waveguide that can support more than one mode of electromagnetic wave propagation.

multipath fading
deflection of radio signal off of obstacles, such as buildings, which can cause interference during signal reception.

MULtiplex-DEultipleX, MULDEX
PCM equipment with analog voice on one side, coded binary signals on the other; also termed "muldem".

multiple
refers to a circuit (including a wire pair) available at several points. It is the basis for multiple outside plant. A multiple, or multipling, is

not desirable because it means that several customers at widely separate living units have access to the same pair. Current practice is to assign each pair to only one customer. Multipling of pairs creates excessive work force activity in the distribution plant because installers have to disconnect a customer at one terminal and go to another terminal to connect a new customer to the wire pair.

multiple pairs

cable pairs that are available to several customers or locations.

multiple plant

plant containing cable pairs that are connected and administered so that when service is discontinued at one location, the pairs can be reassigned for service at another location.

multiple wire

individually insulated wires spirally wrapped together.

multiplexing

simultaneous transmission of two or more signals on the same radio frequency; also see multiplexor.

MUltipleXor, -er, MUX

device that splits and then recombines information signals; it compresses cables by increasing the speed of the signals and pulsing them. It divides a composite signal along a high-speed transmission lines among several channels and then recombines them into one single channel. This strategy allows multiple users to communicate over a single high-speed line and therefore optimizes transmission efficiency.

multiplexor channel

a channel which may do multiplexing by interdigitating bits or bytes; it is designed to operate simultaneously with a number of devices.

multipling

see, multiple.

multipoint connection

a communications line connecting several terminals in different locations.

multipurpose cadastre

a land information system often based on state plane coordinates and containing information concerning property ownership and related materials.

multisensor image

combination of images from different sensing devices, registered to make cells match and produce a sensible final image.

multispectral images

optically acquired images from more than one wavelength interval. See Plate III of San Francisco Bay.

multispectral scanner

remote sensing device--records data in ultraviolet, visible and infrared parts of the electromagnetic spectrum, often from more than one wavelength band simultaneously.

multitasking

capability of computer to perform multiple tasks simultaneously.

multitemporal images

images registered with each other that are accumulated at different times from the same device.

multivariable image

image stored on multiple devices (all registered with each other for proper matching).

multivariate analysis

statistical tools that use multidimensional character of a data to enhance analysis.

Munsell color system

method for designating a color--match it with a pigment color using perceptual dimensions of hue, value, and chroma (intensity).

MUX

MUltipleXor - see multiplex(or).

N n

N connector

a threaded connector for coaxial cable--the N is for Paul Neill.

N2 count

count for allowable number of retransmissions in an X.25 packet-switched networks.

NA

see Numerical Aperture.

nadir

point on the ground directly under the lens of a camera in an aircraft; more generally, from a point, P, on the celestial sphere, any point directly below P (on a ray--directed toward the sphere's center--orthogonal to the line tangent to the sphere at P).

NAK

Negative AcKnowledgment.

nano

metric prefix for one billionth (10^{-9}).

nanometer, nm

one billionth of a meter = 10^{-9} meters. Often used to designate wavelength of light. One nanometer = 10 angstroms. Millicron is another designation for the same unit.

NAPLPS

see North American Presentation Level Protocol Syntax.

NAPP aerial photographs

see National Aerial Photography Program aerial photographs.

NASA

see National Aeronautics and Space Administration.

National Aerial Photography Program, aerial photographs

to date, there are two phases of this program: Phase 1-- leaf off; Phase 2 (in progress)--leaf on.

National Aeronautics and Space Administration

federal agency in the U.S. responsible for developing many of the satellite programs used in remote sensing--such as Landsat.

National Bureau of Standards, NBS

see National Institute of STandards.

National Center for Geographic Information and Analysis, NCGIA

established in 1988 with funding from the National Science Foundation, this set of university research groups at the University of California--Santa Barbara, State University of New York--Buffalo, and the University of Maine--Orono, has been instrumental in evaluating and promoting the interaction of GIS, in a research and teaching context, in the larger university community in the United States.

National High Altitude Program, NHAP

a federally sponsored (USGS) aerial photography program.

National Institute of STandards, NIST

agency that produces Federal Information Processing Standards (FIPS) for all U.S.A. government agencies except the Department of Defense. Formerly National Bureau of Standards.

National Telecommunications and Information Agency, NTIA

a division of the U. S. Department of Commerce.

natural language

computer programming language closer than most to human language.

NBS

see National Bureau of Standards; National Institute of Standards.

NCGIA

see National Center for Geographic Information and Analysis.

ND

see Normalized Difference.

near-infrared

electromagnetic spectral region emphasizing radiation from biomass; typically shorter wavelengths of the infrared region. See Plate III.

nearest neighbor analysis

in a GIS context, a way at looking at nearest cells; suited to data that is not continuous, such as land use, because cell values are dynamically updated and so cannot show gradations in change. Buffers, such as in Plate X, rest on the idea of nearness.

neat line

a demarcation line, drawn on a map, which describes the limits of a specific area, in precise measurements, of the earth's surface and is the common imaginary link to an adjacent map. See Figures 1-24.

neatline point

an interpolated elevation added to a topographic map along its neatline. Some software may have this capability to enhance the topographic expression of features along the map neatline.

negative

reverse of positive; on a photographic image, the dark and light tones are reversed. Numbers on the left half of the number line; those less than zero.

neighborhood analysis

quantitative method in a GIS for assigning a value to a cell based on the values of surrounding cells.

NETBIOS

see NETwork Basic Input/Output System.

network

a mathematical structure composed of nodes and edges linking the nodes; the edges represent flows from one set of nodes to another and often have arrows indicating the direction of flow attached to the edges.

A data structure, used as a structural model, of interconnected linear map features, such as rail lines.

A system of connected computer terminals or data communications facilities. In packet-switched networks, standard facilities are either essential facilities--available on all networks--or additional facilities, available on selected networks. Equipment can be shared and information can be interchanged. Plate XI analyses (visually) a set of flow rates through a road network.

network analysis

quantitative tools for examining network structure which might include, but are not limited to, finding the shortest paths through networks, finding optimal routes, and so forth. There are a number of famous mathematical problems associated with networks, including the Traveling Salesman Problem and the Steiner Problem.

Can also refer to any number of computer systems used for quantitative analysis of geographic relationships that depend on network connectivity, such as flows through the circuits of utility networks. Also see graph theory.

network architecture

the conceptual description of the way communication is accomplished between data-processing gear at disparate sites; it also specifies the processors and terminals, protocols, and software that must be used.

NETwork Basic Input/Output System, NETBIOS

IBM/Sytek-developed software used to link hardware with the NOS or to open session layer communications.

Network File Server, NFS

a distributed file and peripheral sharing protocol developed by Sun Microsystems. Also a version available for PC.

Network Interface Card , NIC

hardware used to physically link a workstation to a LAN's wiring media; firmware and software complete the link to the NOS.

network layer

third level in OSI reference model; its switching, routing, and addressing rules are critical to the operation of large internetworks.

Network Operating System, NOS

system software that creates the network and controls who can use it and how; it can be server-based or peer-to-peer (where every node can be a server of sorts).

network printer

printer shared on a computer network.

network protocol

fixed set of rules used to specify the format of an exchange of data.

network topology

generally, in GIS and related terminology, refers to the connection pattern of communications lines and terminals, or other devices, in a computer network. Also see entry under topology. The network connection pattern in the layer of spiders/stars in Plate VIII is vastly different from the connection pattern in the street layer.

neural network

an array of electrical components having complex input/output properties similar to biological nerve cells; processing power and computer memory are totally distributed across the network (an extension of parallel processing) but connections between sites function efficiently. From a conceptual standpoint, large neural networks might exhibit some of the flexibility, judgmental capabilities and learning abilities of human intelligence. Thus, one might consider that a neural network might be important in scanning complicated hierarchies of layered data, in which the scanner cannot recognize the context in which similar symbols occur. For, a neural network in which the computer learns to make distinctions based on environmental (or other) contexts might serve as a basis for getting scanned images in vector format into a GIS directly from existing paper maps. Digitizing, to get data in a GIS from such a situation, is a cumbersome task.

neutral current loop
a single current loop.

NFS
see Network File Server.

NHAP aerial photographs
see National High Altitude Program.

nibble
one half of a byte (4-bits)--first or last half, only; not a section from the middle.

NIC
see Network Interface Card.

NIST
see National Institute of STandards.

nm
see nanometer.

node
any point in a network--the point might represent any station, terminal, or communications processor in a computer network. In specific pieces of software, the term node may have a meaning specific to that software; in most GIS environments, it refers to an actual intersection point--as a node in a graph. Abstractly, an element of a graph, composed of nodes (sometimes vertices--singular is vertex) and edges. Graphs serve well as structural models for any real-world system composed of a set of discrete entities and a set of flows linking some or all of these entities. Plate VII shows bus routes as nodes linked by edges.

node match tolerance
minimum distance within which two distinct nodes will be replaced by a single node.

noise
random electrical signals that introduce errors and corrupt data. These signals may be generated by natural disturbances of by circuit components.

nominal image
map that approximates an accurate map.

nominal scale
measuring scale along which elements are distinguished by type (name); urban and agricultural are two different types. The lower right corner of Plate II exhibits classification according to a nominal scale.

non-interlaced monitor
a CRT on which the screen refreshing is performed sequentially, moving successively from one horizontal line to the next, from top to bottom.

non-parametric statistics
see parameter.

non-persistent
stations in a Local Area Network that do not try to retransmit immediately following a collision within the network.

Non-Return to Zero, NRZ
any signal which can change level or state directly without having to pass through or return to zero level during the transition.

Non-Return to Zero, Inverted, NRZI
binary encoding method in which a change in state of a signal is a binary 0 and no change is a binary 1.

nonalgorithmic digital records systems
computerized systems that operate on raster databases.

nongraphic data
see tabular data; text or attribute data.

nonsolid color
color formed from a pattern of dots of different colors.

nonvolatile
an adjective used to describe a data storage device that keeps its content even when power is lost.

normal distribution
statistical frequency distribution, symmetric about the mean, bell-shaped and concave down between inflection points at plus or minus one standard deviation from the mean; outside the inflection points it is concave up, monotonically decreasing, and asymptotic on both ends to the x-axis.

Normalized Difference, ND
vegetation indices developed for use with LANDSAT images.

normalized histogram
histogram altered to fit, as closely as is technically possible, a normal distribution (bell curve).

North American Presentation Level Protocol Syntax, NAPLPS
a terminal display protocol.

northing
measurement of Cartesian coordinate distance north a location is from an east-west reference line (such as the Equator); false northing is an adjustment constant to eliminate negative numbers.

NOS
see Network Operating System.

NRZ
see Non-Return to Zero.

NRZI
see Non-Return to Zero, Inverted.

NTIA
see National Telecommunications and Information Agency.

nugget variance
spatially uncorrelated noise in Kriging.

null character
an idle character used to create time for a printer's mechanical actions, such as carriage return.

null modem
see modem eliminator.

Numerical Aperture, NA
a number which defines the light gathering power (angular acceptance) of an optical fiber; the numerical aperture is equal to the sine of the maximum acceptance angle (1/2 the angle of the total acceptance cone).

numerical map
see digital map.

numerical taxonomy
quantitative methods to classify data based on computed estimates of similarity.

O o

object
digital representation of part of the Earth or other entity; also a module in object-oriented programming containing data and procedures. San Francisco Bay is an object in Plate III.

object code
program that a compiler has translated into machine-readable code.

object language
target machine language to which a compiler converts a source program.

object-oriented file system

a file system, based on object-oriented programming, that allows permanent storage of objects and associated links.

object-oriented programming system, OOPS

a programming methodology in which every element in a program is self-contained, having within itself all the data and instructions that operate on that data that are appropriate for that object; one element sends a message to another, and the recipient carries out the task for itself.

oblique photograph

a photograph taken with the camera axis intentionally directed between the horizontal and the vertical. See Plate III.

OC-1

Optical Carrier-level rate of 51.840 Mb/s.

OC-12

Optical Carrier-level rate of 622.080 Mb/s.

OC-3

Optical Carrier-level rate of 155.52 Mb/s.

OCR

see Optical Character Recognition.

octal code

a code based on a number system with a base or radix of 8; a digital system with eight states, labelled 0 to 7, inclusive.

octree

raster data structure to optimize storage requirements for 3-dimensional data (eight is 2 to the third power).

ODYSSEY

polygon overlay program developed at Harvard laboratory for computer graphics.

off-line

transmission of information that does not occur during processing; processing not directly controlled by the CPU.

off-nadir (photograph, viewing)

see oblique photograph.

Ohm

the electrical unit of resistance; symbolized by R.

Ohm's law

stated E=IR, I=E/R or R=E/I, the current I in a circuit is directly proportional to the voltage E, and inversely proportional to the resistance R.

on-line processing

data processing technique in which data sets are input directly from an origin and output directly at the point of use, during processing; processing controlled directly by the CPU.

O(n)

see algorithm.

O(n²)

see algorithm.

ONC

see Operational Navigation Charts.

OOPS

see Object Oriented Programming System.

opaque color

color that masks any other color it is superimposed upon; it does not allow light rays to pass through it.

open architecture

a computer architecture that is compatible with software and hardware from a variety of vendors.

Open Look

GUI interface for X window system developed by AT&T and Sun Microsystems.

open reel tape

9-inch or 12-inch reels of magnetic tape, often used to store medical and satellite images.

Open Software Foundation, OSF

a consortium formed to standardize the UNIX operating system.

Open Systems Interconnection, OSI

a seven layer hierarchical reference interface and communications model sponsored by ISO was formalized in 1984 with the publication of the "OSI Reference Model."   Layer 7 - applications; Layer 6 - presentations; Layer 5 - session; Layer 4 - transport; Layer 3 - network; Layer 2 - data link; Layer 1 - physical.

open window negative

see data mask.

open wire

single, non-insulated wire used in rural areas.

operating system

a collection of software and firmware utility programs that guide a computer in performing its most basic functions, such as allocating memory and disk space and controlling various input/output operations.

Operational Navigation Charts, ONC

U.S. DMA 1:1 million scale chart series--charts used as sources for DCW among many other general uses.

operator

algebraic term whose interpretation may vary when viewed in software contexts; but, traditionally, a descriptor of an action to occur to that which follows. Thus, in the Newtonian view of the calculus, D serves as an operator to indicate that the expression which follows immediately is to be differentiated.

optical character recognition, OCR

the automatic recognition and interpretation of text.

optical directional coupler

a coupler for optical fibers in which lightwave propagation may be in one or both directions. Often used in optical fiber communications systems, such as CATV, to combine or split optical signals.

optical disk

a storage device on which several gigabytes (typically up to 500 megabytes per side) of data can be stored. A laser is used to make a permanent record of the data on the disk in such a way that any portion of the disk can be written on only once although it can of course be read many times--WORM--Write Once, Read Many. There are also Erasable Optical (EO) disks and Magneto Optical (MO) disks.

optical fiber

a single optical transmission waveguide formed from a (glass) fiber core and a fiber cladding that can guide a lightwave. Each glass fiber is an independent circuit. The fiber is generally cylindrical in shape and its axial length is large when compared to its diameter. Internal reflection is used to transmit light along the axial length of the fiber. The amount of light lost on the journey from one end of the fiber to the other depends on a variety of factors having to do with the physics of light passing through a glass-like medium.

optical fiber preform

optical fiber material from which an optical fiber is to be made; the actual fiber is generally much longer than the preform from which is was created. For example, a heated glass cylindrical preform might be drawn out as molten glass to become a slender fiber. A standard 1 meter preform will yield about 12 kilometers of fiber.

optical scanner

a peripheral that scans a document (graphics, text, and symbols) optically, sensing variations in reflected light patterns, and encodes the information in a digital file from which raster hard copy can be produced

optical waveguide

a transparent filament of high refractive index core and low refractive index cladding that transmits light.

optimal estimator
quantitative tool to minimize value of a given criterion function.
optimal path selection
use of network database to find an optimal path based on various constraints.
optimization
generally, a set of quantitative procedures for making something as good as possible within various parameters; in a GIS display, often refers to choices for color assignment from a large set of possibilities to produce an harmonious product giving good fidelity of color in natural scenes. On Plate VIII, the stars overlap at only a few places. In an optimal arrangement of stars, a boundary may easily be drawn to separate the stars and assign each to a separate compartment of space. When stars intersect, this strategy is not possible. Thus, any map of this sort in which stars intersect depicts a non-optimal arrangement. Dirichlet regions are an example of an optimal partitioning of a plane space.
ordinal
measuring scale on which units are distinguished by rank derived from a quantitative measure. Plate I, right side, illustrates an ordinal ranking by color.
ordination
reduction of complex data sets to make them more comprehensible; principal component analysis is one quantitative tool that fits this category.
orientation
direction; might be north, south, east, or west; might be clockwise or counterclockwise (with reference to a clock with hands); or, it might be any of a variety of others. On Plate IV, the base map provides orientation; without it, the data--represented as disks of varying diameter, would make no sense.
orphan
an entity occurrence without a parent entity, e. g., a complement without a parent entity (cable).
orthogonal
perpendicular; right-angled; at right angles to.
orthographic perspective
three dimensional display surface formed from a dense set of parallel profiles. See Plate IX.
orthophotograph
aerial photos that have been manipulated so as to reduce distortions from camera tilt and terrain relief; often they are made from very high resolution stereo pairs of aerial photos.

orthovideo
> airvideo images that have had distortion removed, often by aligning points in the photo with control points.

oscillation
> regular fluctuation, as in the motion of a pendulum, a sine wave, or a pattern of numbers such as {0, 1, 0, 1, 0, 1, 0, 1,...}.

OSF
> see Open Software Foundation.

OSI
> see Open Systems Interconnection.

OSP
> see OutSide Plant.

out cable
> in dedicated outside plant, the originating cable (usually distribution) that leaves a control point or access point. Under the serving area concept, a distribution cable that is designated for use solely within a serving area and which terminates in and originates at the serving area connector.

out distribution pairs
> in the serving area concept, pairs designated for use solely within a serving area. Out distribution pairs originate at the serving area interface. Current practice is to plan for at least two distribution pairs for every living unit ultimately forecast for the serving area.

output
> result of processing data--in a GIS or in other software.

OutSide Plant, OSP
> all of the telephone hardware used to connect subscribers to switching offices, and switching offices to one another. Outside plant includes underground conduit, underground cable pole line, aerial cable, open wire, buried cable, submarine cable, building cable, wire, distribution terminals, splices, and distribution interfaces.

outside plant location
> records that show placement and attributes of outside plant (e.g., distribution terminals and location, length, size, date, and nomenclature of cables, etc.).

outside plant map
> a geographic baseline map showing location and structure type for the various portions of the cable plant in the network--of the sort shown in Plate V.

overedge
> part of the map lying outside nominal border.

overhead bit
  a non-data bit used in addressing, error control and in a variety of other situations.

overlap
  the amount by which one photograph covers the same area as covered by another, customarily expressed as a percentage.

overlay
  layer superimposed on another layer of a map (or other image); often with registration marks to ensure precise alignment when layers are combined to make a new map. Plate VIII shows a layer of spiders/stars as an overlay on a street grid; Plate VII shows a layer of bus routes, instead, as an overlay on the same grid. Also see Plate I.

overlay analysis
  technique for combining spatial information from a number of maps to create a new map with new boundaries. Even within one map, new smaller pieces may be created in this manner. For example, in Plate IV some small circles (symbols) are overlaid on larger ones; when they are, a small rim of darker color is left so that the smaller symbols are not engulfed in the larger symbols--were they, accurate analysis would be inhibited.

overprint
  text or data printed on a map that is in addition to the original material.

oversampling
  a technique in which each bit from each channel is sampled more than once.

overshoot
  an edge that reaches beyond one node but does not extend to reach another node.

overspeed
  situation in which transmitting device runs faster than the data offered for transmission; an overspeed of 0.1% for a modem is common.

Ozalid process
  see blue-line.

P p

P-N junction
  see, Positive-Negative junction.

p-persistent
  describes situation in which stations along a Carrier Sense Multiple Access LAN try to retransmit, with probability p, soon after a collision.

PABX
    see Private Automatic Branch eXchange.

PAC
    see Plant Account Code.

package
    set of computer programs.

packet
    a group of information transmitted over a network; each packet includes addressing and control data.

Packet Assembler/Disassembler, PAD
    a device used to convert between terminal protocol(s) and X.25; it may be single or multiple channel, asynchronous or synchronous.

Packet Data Network, PDN
    might refer to either a Packet-Switched Network or to a Public Data Network.

packet header
    the first three octets of an X.25 packet in a packet-switched network.

Packet Switched Data Network, PSDN
    an X.25 protocol communications network.

packet type identifier
    the third octet in the packet header identifying the packet's function and sequence number.

packet-switched network
    data communications networks, such as Local Area Networks and Public Data Networks, that transmit packets. Packets are sent from origin to destination via virtual circuits.

PAD
    see Packet Assembler/Disassembler.

page layout
    design of a page, including text, graphics, margins, and fonts.

paint
    on a CRT, to fill an area with color or other symbolism.

pair
    see cable pair.

pair group
    all cable pairs that are part of a feeder route that serve an allocation area.

Pair-Gain Systems, PGS
    electronic systems that allow a smaller amount of physical wire pairs to serve a larger number of customers.

pairs terminated
    feeder pairs terminated on a main distributing frame.

PAL

see Programmable Array Logic.

palette

as with an artist, in a GIS setting this refers to the combination of colors that can be displayed simultaneously; the set of colors from which the artist can choose.

PAM

see Pulse Amplitude Modulation.

PAMA

see Pulse Address Multiple Access.

panchromatic image

often refers to a black and white photograph of a color scene, even though the image may be collected in a broader visual wavelength range using film that is sensitive to all of the visible spectrum of light.

panelled cell

the completed assembly of pieces of film positives onto a grid or projection which is used as a base for DTED.

parallax

apparent change in position of separated locations when seen from different vantage points: for example, the base and top of a water tower seen from different positions along the flight path of an airplane flying over the water tower.

parallel

circles used to partition the Earth's surface; they are parallel to the Equator which is the only one of this set which is a great circle. All others are small circles (often equally spaced) parallel to the Equator. The extent to which a small circle is north or south of the equator is measured in degrees of latitude, with values ranging from 0 to 90 degrees.

parallel communication

see parallel transmission.

parallel port (interface)

connection between a computer and a peripheral such as a printer; this connection will use parallel transmission.

parallel processing

computer architecture which involves using multiple processors within a single computer; as with parallel transmission, this sort of parallel strategy leads to a structure which can very effectively process much information.   Single processor systems process data in a sequential manner.

parallel transmission

a transmission technique that sends each bit simultaneously over separate lines. Often used to send data a byte (=8 bits) at a time over 8 lines to a high-speed printer--as contrasted with serial transmission. While parallel transmission is very effective, it presently can be used only over limited physical distances of several feet or yards.

parallelepiped classifier

see boxcar classifier.

parameter

arbitrary constant each value of which characterizes a member of a system; also, a number or value that describes a statistical population, such as mean (average value of the distribution), median (halfway point above which half the values in the distribution lie), or mode (most frequent value in a distribution). That which is "nonparametric" might thus refer to criteria other than descriptive values for a statistical population.

parametric map

map with criteria that, taken together, describe a statistical population.

parcel

an area of land, often defined by legal boundaries. In the United States, these boundaries are often described using the terminology of the United States Public Land Survey System of townships, ranges, sections, and multiples of two subdivisions of these units (often powers of two, as halves, quarters, eighths, sixteenths, or thirty-seconds of a section, or even perfect squares, as 36 sections in a township). Township and range lines serve as a coordinate system for the Public Land Survey. See Plate II, upper right; Plate V.

Parcel Identification Number, PIN

numbering system for identifying parcels of land that are part of a database in a computer system; numbering strategies might include mere rank-ordering or a more complicated geocode that shows parcel location as well as numerical order.

parity

number of 1s in a group of bits; parity is thus either even or odd, depending on whether the number of 1s is even or odd.

parity bit

the bit which is set to 1 or 0 in a character in order to guarantee that parity is even or odd.

parity check

addition of overhead bits to ensure that parity is always either even or odd. This strategy allows detection of single transmission errors.

partition

logical or physical segment of disk space in a computer; a hard drive might be logically or physically partitioned into two parts, one to be assigned to one set of users and the other to a different set of users.

PASCAL

high-level programming language; often used in scientific programming; excellent language for teaching structured programming.

pass point

a point located on two rasters that cover overlapping geographic territory. Useful in tying multiple images together. See Plate II.

passive system

remote sensing system that measures emitted radiation.

password

a protective word associated with a user name; a user logging into a computer system must supply the correct password before the system will permit access.

patch method

methods of surface fitting, much like creating a tortoise shell. Patches are pieced together to try to form a smooth continuous surface. See Plate II.

path

DOS path describes chain of directories, subdirectories, and so forth to pinpoint location of a file in the computer.

pattern matching

process of overlaying multiple images in order to determine the frequency of occurrence of a particular pattern.

PBX

see Private Branch eXchange.

PC

see Personal Computer.

PC648

obsolete--a type of above ground distribution terminal; treat as DSB.

PCIPS

raster microcomputer image processing system.

PCM [1]

see Plug Compatible Manufacturer.

PCM [2]

see Pulse Code Modulation.

PCN

see Personal Communications Network.

PCP

see Preferred Count Primary pairs

PCS

see Permanent Secondary Pairs; Personal Communications Service; Preferred Count Secondary Pairs.

PD

see PhotoDetector.

PDES
see Produce Data Exchange Specification.
PDN [1]
see Packet Data Network.
PDN [2]
see Public Data Network.
pedestal
an above ground metallic enclosure in which the cable loops in and out again with access to any of the pairs.   Used with buried cable. Might be found on maps such as Plate V.
pel
another word for picture element.
pen plotter
see, plotter.
performance
extent to which a computer system meets specifications.
perimeter
boundary of a polygon.
peripheral
any physical (hardware) device separate from but connected to and controlled by the central processing unit of a computer: printers, plotters, scanners, or modems.
Permanent Secondary Pairs, PCS
distribution pairs for second-line residential service that are permanently committed from a serving area interface to a housing unit. In new plant, these pairs are pre-assigned by the engineering forces. At the time of cable installation, they are terminated in the interface and spliced to a buried service wire and may be cut off beyond their service connection point by the construction forces. In existing plant, they are assigned by the plant service center forces and are not cut off beyond the service connection point unless specified by an engineering work order.   When engineering predicts that more than two pairs will be required per housing unit, all additional pairs pre-assigned for the third, fourth, or more residential requirements will be permanent secondary distribution pairs.
Permanent Virtual Circuit, PVC
a circuit between two users in a packet-switched network on which no call setup or clearing procedures are necessary.
permanently connected plant
plant containing dedicated (permanently committed) cable pairs.
perpendicular
orthogonal; vertical; upright; at an angle of 90 degrees.

persistent

in a Carrier Sense Multiple Access LAN, a term to describe the situation in which stations try to retransmit almost immediately after a collision.

Personal Communications Network, PCN

digital wireless radiotelephone network employing low-powered (mobile) wireless phones, data terminals, and microcell base stations joined to the switched, public telephone network.

Personal Communications Service, PCS

FCC term to describe intelligent, digital wireless, personal two-way communications system.

Personal Computer, PC

a microcomputer, used in professional situation and in home computing, for a variety of applications that are end-user oriented.

perspective view

three-dimensional block diagram, generated in a GIS from a digital elevation model.

PEX

an extension of PHIGS to X Windows, PEX allows programs incorporating PHIGS to run in windows on different types of terminal and workstation displays.

PGS

see Pair-Gain Systems.

phase modulation

a way to add information to a sine wave signal by modifying the phase of the wave to match characteristics of the information to be transmitted.

Phase Shift Keying, PSK

a phase modulation technique.

phi

Greek letter conventionally used to denote latitude as the first component in an n-tuple describing geographic location.

PHIGS

see Programmer's Hierarchical Interactive Graphics System.

photo base

distance, on adjacent prints of vertical aerial photos, between principal points.

photo interpretation

analysis of portions of the Earth's surface from aerial photographs. One can easily note variations in elevation in Plate III.

PhotoDetector, PD

a light detector that can extract information from an optical carrier.

photodetector responsivity
    an index, calculated as a ratio of the root mean square of the voltage over the root mean square of the incident optical power input, may be used to measure the responsivity of a photodetector.

photodiode
    a diode whose conductivity increases directly with increased incident electromagnetic radiation.

photogrammetric digitizing
    use of photogrammetric instruments, such as an analytical stereoplotter rather than a table digitizer, to compile new maps from aerial photographs.

photogrammetric mapping
    use of photogrammetric techniques to compile maps from measuring controlled aerial photos.

photogrammetry
    use of aerial photographs to make decisions and reliable measurements for surveying and mapping.   Controlled aerial photographs are compiled to make new maps, often using a photogrammetric digitizer. This latter device is similar to a manual digitizer in which the table is replaced by photogrammetric equipment. The result is a map of high precision of the features recorded by the photogrammetric process.

photographic interpretation
    analysis of photographs in order to identify objects in the photo and consider their significance and relation to each other in space.

photographic map
    map made from assembling photographs and adding appropriate information such as a reference system, grid, labels, scale, and so forth.

photolithography
    conversion of a photo into an image on a lithographic plate.

photon detector
    detector used in remote sensing.

physical layer
    first level in OSI reference model; governs physical network connection, including connectors, timing, and voltages, of workstations to the wiring media.

physiographic
    description of physical characteristics of a site. See Plate II.

PIC
    see Plastic Insulated Conductor

pico
    prefix for one trillionth (10 to the power of -12).

picture element
  see pixel.

PIN
  see Parcel Identification Number.

pin diagram
  a type of three-dimensional diagram formed by drawing straight pins, vertical lines topped with a node, at various locations on a flat map. The pins represent the z-coordinate in a Cartesian characterization of three-dimensional space. A pin diagram can be used to offer a discrete view of topography much as Plate IX offers a continuous view.

PIN diode
  see Positive-Intrinsic-Negative diode.

pins, registration
  used to hold paper or plastic map layers in place so that they might be registered accurately.

pit
  depression in a surface, including in a digital surface as a consequence of noise. See Plate IX.

pitch
  a rotation of an aircraft about the horizontal axis-up or nose-down attitude.

pixel
  picture element - the basic data unit of a raster system--the smallest element of single uniform density in a photographic or other image. The physical dimensions of pixels are determined by resolution and map scale. Abstractly equivalent to a cell in a mathematical matrix.

pixel depth
  see color depth.

Plain Old Telephone Service, POTS
  usually includes push-button dialing, nationwide and international direct dialing, and regular billing.

plan position indicator
  radar system with circular screen and sweep hand that indicates the position on the screen of returning signals.

plane coordinates
  generally, coordinates for a plane; often used to refer to a plane that is developed from a conical or cylindrical surface. The orthogonal axes in the latter case are eastings (measured along the x-axis) and northings (measured along the y-axis).

plane of best fit
  plane fit to a data set in three dimensions, using least squares.

planimeter
   mechanical tool to measure areas in the plane; a wheel on a rod with a mark on the wheel to count revolutions.

planimetric
   map of horizontal position only; depicts features on the earth's surface that are visible and identifiable from aircraft and aerial photos. When surface features can be captured in this manner, they can be mapped using photogrammetric or surveying procedures as a planimetric map.

planimetry
   study of horizontal control; all vertical relief is omitted.

plant
   a general term applied to any of the physical property of an operating company that contributes to the furnishing of communications services. See also item of plant. Might be displayed on a map such as Plate V.

Plant Account Code, PAC
   see field reporting code.

Plant Location Record, PLR
   records showing the position and/or method of placement of telephone outside plant.

Plastic Insulated Conductor, PIC
   a type of multiple pair cable that is insulated with various types of polyethylene. PIC cable is color coded for easy identification.

plat map
   map showing property and political boundaries; similar in appearance to Plate V.

platform
   workstation platform often used as the host for a GIS or other mapping software.

Plot Request Form, PRF
   the request used to produce a drawing on the plotter.

plotter
   a peripheral that renders graphic plots of computer data. Plotters are typically capable of generating large-format maps using vector files to generate the line elements of the map so that curved lines appear as a clear, crisp image.

plotter font
   scalable font created as a sequence of dots connected by lines.

PLR
   see Plant Location Record.

PLSS
   see Public Land Survey System.

Plug Compatible Manufacturer, PCM

Plug-Compatible Machine. Term used about a machine to describe one that can be substituted directly for a manufacturer's original equipment, presumably as an improvement over the original.

PM

see Prime Meridian.

point

in a digital mapping setting, a zero-dimensional entity representing a location as a coordinate pair (or n-tuple in higher dimensions). Oil wells or lighthouses are examples of points in this context.

point data

data consisting of single (x,y) coordinates, in a vector data structure; data consisting of a single cell in a raster data structure.

point in polygon

capability to determine the intersection of two overlays, one of points and one of polygons, in order to assess which polygons contain points.

point size

in typesetting, there are 72 points to an inch. In many word processors, one has the capability to select spacing in points. Font height is often expressed in points.

point-to-point connection

same as link.

Point-to-Point Protocol, PPP

a variation of the serial line internet protocol that minimizes the amount of data sent by removing redundant protocol information from sequential data packets.

pointer

computer tool to choose a graphic element.

polar transmission

see bipolar transmission.

polarization

orientation (north-south, as in compass point orientation) of vibration of electrical field vector of radiation (electromagnetic).

pole

a column, usually made of wood, used to support aerial structures such as telephone cable.

pole to pole guy

a strong steel wire that runs from one pole to another for support.

poll

a message sent by a master computer, querying each station along a network as to whether or not it has anything to send across the network.

polyconic projection
  see conic projection.
polygon
  on a CRT, a closed figure (shape) of three sides or more bounded by line strings intersecting at nodes. Numerous theorems about polygons in two dimensional space (or higher dimensions) have corresponding counterparts in the Manhattan space of the CRT with square pixel units; many, however pose difficult problems. The ideas of convex and concave polygon are clear in both the Euclidean and the CRT plane; problems can arise when the two-dimensional character of the fundamental unit (the pixel) of the CRT replaces the zero-dimensional character of the fundamental unit (the point) of the plane. Theorems seem to carry through as long as the concept of point and pixel are viewed as interchangeable at the abstract level; there are similar problems, often greatly magnified, in higher dimensions. See also, Jordan Curve Theorem; Voxel. Counties on Plate X are polygons.
polygon data
  in a raster data structure, a set of contiguous cells of the same value; in a vector data structure, data points enclosed by a perimeter of lines.
polygon dissolve
  dissolution of contiguous polygons by removing shared boundaries.
polygon merge
  merging of contiguous polygons by removing shared boundaries.
polygon overlay
  generation of new polygons for a map from GIS layers containing previously established polygons; often, the generation relies on set-theoretic union, intersection, and subtraction. The buffers in Plate X may be viewed as a union of an infinite number of circles (from an unseen layer) centered on each point of the interstates. Also see overlay.
polygonization
  connecting otherwise separate arcs to form polygons.
polynomial
  mathematical expression of the form

$$a_0x^0 + a_1x^1 + ... + a_nx^n,$$

where n is an integer. The individual parts linked by plus signs are called terms. Within each term is a variable (x in this example) raised to an integral power (from 0 to n) and a coefficient (a's here). There are always a finite number of terms; when n is small, there are specialized names for polynomials that tell exactly how many terms the polynomial under consideration has. Thus, a binomial is a polynomial of two terms; a monomial is a polynomial of one term; and, a trinomial is a polynomial of three terms.

popup; pop-up program

information that pops up on the computer screen when a function key is pressed; memory-resident program.

port

the name of the socket, connector, or computer interface at the back of the computer to which a terminal, printer, or other communication device is connected.

port concentrator

equipment that allows several terminals to share the same computer port.

port selector

see data PABX.

portability

in a computer environment, the capability to easily move software from one computer to another.

porting

modifying code that runs on one hardware platform or operating system so that it will properly execute on another hardware platform or operating system.

positive

image with light and dark tones matching those of the object they represent. Numbers on the right half of the number line; those greater than zero.

Positive-Intrinsic-Negative diode, PIN diode

a semiconductor photo sensitive cell which has a large intrinsic (very lightly doped) region sandwiched between Positive and Negative doped semiconductors; photons absorbed in this region release electrons that are attracted by the electric field to generate a current. Its conductivity changes in accordance with incident light.

Positive-Negative junction, P-N junction

the dividing line in a semiconductor between a P region and an N region; electrons can flow from N to P but not from P to N.

POST

see Power-On System Test.

Post, Telephone, and Telegraph authority, PTT

governmental agency found in various countries that functions as, and administers, the communications common carrier.

posting

the transfer of information from completed work orders to the Plant Location Records.

PostScript
page-description computer language of Adobe Systems, Inc. Particularly useful in computerized typesetting applications and desktop publishing with graphics.

potential
estimate of possible interaction between separated locations.

POTS
see Plain Old Telephone Service.

pour point
in watershed analysis, the point where water from one watershed would pass into the adjacent watershed. See Plate IX.

Power-On System Test, POST
set of computer start-up test routines.

PP
see Pre-Posting.

PPP
see Point-to-Point Protocol.

Pre-Posting, PP
transferring information from the work order to plant location records prior to completion of a construction project.

pre-processing
organization of data set to make it manageable and susceptible of further processing.

precision
the degree or exactness with which a quantity is stated; a relative term often based on the number of significant digits in a measurement. Sometimes expressed as a variance of repeated measurements-- replicability.

predigital compilation
the selection, assembly and graphic presentation of all relevant information required for the preparation of such information that may be derived from other maps, charts or other topographic sources.

Preferred Count Primary pairs, PCP
feeder pairs terminated on the feeder (IN) field of the serving area connector and designated for interconnection to primary distribution pairs. Once connected, they will be administered either on a dedicated or a connect-through basis.

Preferred Count Secondary pairs, PCS
feeder pairs terminated on the feeder (IN) field of the serving area connector and designated for interconnection to either permanent or reassignable secondary distribution pairs. These feeder pairs are available for reassignment from the central office to the serving area connector upon disconnection of service.

premises network
    same as cable system.

presentation layer
    sixth level in OSI reference model; reformats differences in user data into a form usable by the network's application layer.

pressurization
    the use of pressurized gas or dry air inside of cable sheaths to prevent the entry of water at faulty splices, damaged sheath, or accidental sheath openings.

PRF
    see Plot Request Form.

PRI
    see Primary Rate Interface.

primary colors
    set of fundamental hues that can be mixed to match any color. Additive primary colors: blue, green, red.
Subtractive primary colors: cyan, magenta, yellow.

primary distribution pairs
    distribution pairs that are permanently committed from the serving area interface to an addressable location.   They are entered as permanent information on the dedicated plant assignment card and are connected to preferred count primary feeder pairs at the serving area interface. In new plant, they are pre-assigned by the engineering forces. At the time of cable installation, they are either terminated on the binding post of a terminal, or they are sliced to a buried service wire and cut off beyond their connection point by the construction forces. In existing plant, they are assigned by the plant center forces and are not cut off beyond their service connection point unless so directed by an engineering work order.

Primary Rate Interface, PRI
    ISDN transmission with 23 "B" (Bureau) channels and 1 "D" (signaling) channel; see BRI.

Prime Meridian, PM
    reference meridian from which longitude is measured (adopted in late nineteenth century); it passes through the Royal Observatory in Greenwich, England.

primitive
    basic concept upon which other, generally more complex, concepts rest. Within an electronic environment, therefore, primitive commands to produce hard copy might go from a printer driver to produce hard copy; what is in the hard copy is more complex and is not part of the simple, primitive, idea of getting a printed copy of whatever the content is.

principal components analysis
    statistical technique to reduce complex sets of data by reducing their dimension and using those components that account for much of the variance as a means to simplify the data or the corresponding image.

principal point
    intersection of lines joining opposing fiducial marks on a photo; center of a photograph.

print queue
    a method of controlling the execution of printing. There are various ways to control the line of jobs.

print server
    a networked computer that lets other workstations print to its attached printers.

printer
    a peripheral that produces alphanumeric hardcopy printouts of computer data.

printer converter
    a coaxial converter permitting an asynchronous printer to emulate another printer.

printer driver
    see driver.

printer font
    fonts sent to the printer by the computer or those stored in the printer's memory.

Private Automatic Branch eXchange, PABX
    a user-owned automatic telephone exchange.

Private Branch eXchange, PBX
    a manual, user-owned telephone exchange.

private line
    same as leased line.

probability
    branch of mathematics that deals with likelihood of events. Interconnected with many statistical measures as a theoretical backdrop.

process colors
    colors used on duplicating equipment to duplicate an original: black, cyan, magenta, yellow.

processing gain
    amount of digital information received as compared to transmission interference.

processing mask
    see data mask.

processing unit
hardware to execute programs and control functions of a computer system.

Produce Data Exchange Specification, PDES
a data exchange format for 2-D and 3-D graphics files.

profile [1]
a vertical cross section of the surface of the ground, map, model, et cetera along a definitely located (fixed) line which portrays its shape, configuration, et cetera.

profile [2]
in a packet-switched network, parameter values for a terminal which, when stored, can be recalled and used as a set by selecting the appropriate profile of values.

program
a precise set of coded instructions and logical statements that guides a computer in the performance of a specific task or tasks.

Programmable Array Logic, PAL
color video standard not compatible with NTSC; in common use in Europe.

Programmable Read-Only Memory, PROM
permanently stored data in a memory (nonvolatile semi-conductor).

Programmer's Hierarchical Interactive Graphics System, PHIGS
a graphics software standard.

programming languages
language in which to write computer programs: FORTRAN, C, PASCAL, and others are examples. See related topics: assembler, compiler, interpreter, fourth generation language, and individual languages.

progression
sequence exhibiting discernible relationship between adjacent terms.

progressive sampling technique
data sampling technique.

projection
mathematical transformation for representing parts of the Earth's surface on a flat sheet of paper; see also names of various map projections. See figures 1-24.

PROM
see Programmable Read-Only Memory.

prompt
a word, phrase, or noise such as a beep, used as a cue to prompt a user response at a CRT terminal.

proof [1]
  verification of a mathematical conjecture through the use of a logical argument based on known facts.

proof [2]
  preliminary copies of a map, photo, or text.

propagation delay
  transit time of a signal through a link or network.

propagation mode
  an electromagnetic field condition that can exist in a fiber optic waveguide; also called transmission mode.

protected mode
  computer operating mode that can directly address extended memory.

protected terminal
  a terminal that has equipment placed in it that prevents foreign voltage from travelling through outside plant facilities.

protocol
  a language and a set of error checking rules and standards used for communications between dissimilar computer or telecommunications systems.

protocol converter
  equipment that translates from one communications protocol to another.

proximity analysis
  technique in a GIS that creates polygons around randomly spaced points; this technique is one way to generate pictures of discontinuous data. In a mathematical environment, this is much like using step functions to generate pictures of discontinuous data--yet another connection to theory.

PSDN
  see Packet Switched Data Network.

pseudo node
  a node with exactly two incident edges--a GIS term. A pseudo node is in fact a graph-theoretic node.

pseudo-color image
  process of assigning colors in such a way as not to produce a "true" image--such as assigning levels of gray to colors, thus obtaining a "fake" color image.

pseudo-cylindrical map projection
  projections designed to minimize the appearance of angular and areal distortion. Tobler's equal area hyperelliptical (base map for Kolars and Nystuen textbook in references section) and Robinson's

compromise projection (Figure 24) are pseudo-cylindrical projections employed in this capacity.

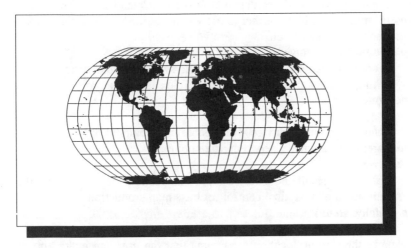

Figure 24. Robinson projection.

pseudo-random
in a GPS context, a signal with noise that appears random but in fact has a regular discernable, although likely complex, pattern of binary digits.

PSK
see Phase Shift Keying.

PTT
see Post, Telephone, and Telegraph authority.

Public Data Network, PDN
a network serving the data communications needs of the public-- might be operated by any of a variety of sources, such as common carriers or private companies.

public key cryptography
a method of encryption of messages in which the method of encryption is revealed publicly and yet in which decryption is only probable by the person revealing the code. One such method is to provide two integers k and n, instructing the encrypter to encrypt a number representing a letter or block of letters (e.g., 190704 might represent THE) by taking that number to the kth power modulo n. If n is a product of 2 large primes, it is difficult to figure out how to decode the message unless the two primes are known. If two people each have public keys revealed, sophisticated use of this system even allows them to send messages which identify the sender.

Public Land Survey System, PLSS

a system for referencing public land in U.S.A. that was laid out during the nation's early settlement. It partitions land areas into 6 mile by 6 mile townships further subdivided into 36 one mile square sections. Also see parcel.

public switched network

a communications system relying on switching that offers circuit switching to a number of customers. Telex is an example of such a network.

puck

hand-held pointing device with visible cross-hairs to aid in accuracy in targeting locations. Used to digitize points.

pugging

drilling a small hole in a film diapositive to assist in the establishment of analytical control for base map production.

pull-down menus

a program interface that uses key words at the top of the screen to identify the starting location for menus that the user can make appear below them.

pulp cable

conductor insulated with wood pulp applied directly on the wires. Pulp cable is not color coded.

pulse

a signal of short duration for the transmission of information; implies very short rise and fall times on the leading and lagging edges of the waveform.

Pulse Address Multiple Access, PAMA

where carriers are distinguished by their temporal and spatial characteristics simultaneously.

Pulse Amplitude Modulation, PAM

a pulse (or stream of pulses) whose amplitude is the instantaneous value of an analog or voice frequency signal; a coding scheme in which pulses are of uniform time duration (width), but vary in amplitude (height) in accordance with the signal amplitude at the time of sampling, one pulse per sample.

Pulse Code Modulation, PCM

a method of digital coding in which the information is represented by the presence or absence of a pulse in a pulse period; converts analog voice signals into digital format.

pulse length

the time duration of the burst of energy emitted by a pulsed laser; also called the pulse width; usually measured at the "half power" points.

pulse slot

space allotted on a time division basis for inclusion or exclusion of the binary information in the form of the presence or absence of a pulse.

purity of color

percent saturation of a color by its dominant wavelength color.

purposive sampling

data sampling method.

push brace

a pole that is placed at an angle to the line pole to provide support. Used in lieu of a guy and anchor.

pushbroom scanning

scanning using progressive lines of ground scan as a sensor moves overhead.

PVC

see Permanent Virtual Circuit.

PVL

Parameter = Value Language.

Q q

Q bit

Qualifier bit; bit number 8 in the first octet of a packet header. Used to indicate whether or not the packet contains control information.

QA

see Quality Assessment.

QAM

see Quadrature Amplitude Modulation.

QIC

see Quarter-Inch Compatibility.

quadrangle

geometrically a polygon determined by four angles; thus, a quadrangle has four vertices, four sides, and four angles. In a mapping context, often refers to the area covered by an apparently rectangular paper map, or it may refer to the geographic area covered by electronically stored data.

Often maps labelled "quadrangles" are in fact isosceles trapezoids; the non-parallel pair of opposite sides are equal segments of meridians (converging eventually at the poles); the other pair of parallel opposite sides run along parallels of latitude.

quadrant

one quarter of a circle.

Quadrature Amplitude Modulation, QAM
technique designed to increase the number of bits per baud by combining phase modulation with amplitude modulation.

quadrilateral
a) geometric figure formed from four (infinite) straight lines: thus, a complete quadrilateral has four sides, six vertices, and six angles.

b) more commonly in elementary geometry, a closed geometric figure formed from four straight line segments any one of which intersects exactly two of the others. A typical quadrilateral thus contains an area interior to it; common examples include square, rectangle, parallelogram, rhombus, and trapezoid.

quadtree
raster data structure used to reduce storage requirements. Contiguous homogeneous areas are coded singly. In a given raster, the set of cells is partitioned successively into four equal non-overlapping, exhaustive, squares. Partitioning continues until each unit is homogeneous or further partitioning is no longer possible. The strategy is abstractly identical to the manner of subdivision employed in the U.S. Survey of Public Lands. When a square area is represented as a node, and the successive partitioning is represented as lines joining nodes, the emerging diagram is a graph-theoretic tree. The quadtree, however, unlike the abstract graph-theoretic tree, is not translation invariant--it matters, using a quadtree, where the nodes are in terms of coordinates.

qualitative
that which is not quantitative; in mapping, text rather than numeric labels of a feature are qualitative. Land use types are often qualitative.

Quality Assessment, QA
set of procedures designed to ensure that a project meets specified accuracy and precision standards.

quantitative
features that can be characterized by real numbers--temperature, elevation, and so forth.

quantization
division of a range of continuous values into a finite number of discrete levels.

Quarter-Inch Compatibility, QIC
an industry standards group and related standards for quarter-inch tape cartridge drive systems.

query
to ask for specific information from a database.

query language
see Structured Query Language, SQL.

queue

a list formed by items waiting in line for service, such as a list of messages waiting for transmission.

QWERTY keyboard

the standard American keyboard layout; named for the first six letters in row two of the keyboard.

R r

"R" field code

repair. Ordinary repairs including the cost of repairing plant damaged by casualty.

rack

same as cabinet.

rack mount

equipment designed to be installed in a cabinet.

RAD50

a computer code (also known as RADIX-50) based on a number system with a base or radix of 50; RAD50 coding allows three characters to be stored in a two byte word.

RADAR

see RAdio Detection And Ranging.

RADAR shadow

area of no return on a radar screen.

radial displacement

image shift caused by difference in elevation between that of the object and that at the image (photo) center.

radially stratified fiber

see step-index fiber.

radiance

energy flow measured in units of power (such as watts).

radiation

process of emitting and propagating electromagnetic (or sound) waves through space.

RAdio Detection And Ranging, RADAR

method using returned radiation signal to locate objects, acquire a terrain image, and execute various tracking functions.

Radio Frequency Interference, RFI

electromagnetic radiation in the radio frequency spectrum from 15 to 100 GHz; the best shielding material for RFI is copper and aluminum alloys.

radio waves

long wavelength radiation in the electromagnetic spectrum.

radiometer
　passive device to measure electromagnetic radiation intensity.
RAM
　see Random Access Memory.
RAM cache
　see cache memory. See also, RAM disk in contrast.
RAM disk
　RAM reconfigured to work beyond 640K. Contrast with RAM cache.
random
　selection process in which each element of a set has equal likelihood of being chosen.
Random Access Memory, RAM
　electronic storage device (volatile semiconductor) into which data can be written and read.
range
　see parcel.
range extraction
　method of converting a gray scale to binary; everything with the a selected interval (range) is assigned the value of 1 (white) and all else outside the range is assigned 0 (black).
raster
　two dimensional array for storing, processing, and displaying graphic data in which graphic images are stored as values for uniform grid cells (pixels). In raster displays, the display is usually processed line-by-line, as opposed to vector graphics displays, which process objects such as circles as a whole.
raster cell
　one value in a raster, located by row and column position, as in a matrix.
raster data
　data displayed as discrete picture elements (pixels).
raster display
　see raster data.
raster to vector conversion
　transformation carrying an image formed of cells to one formed of lines and polygons--raster to vector--also, can be phrased in reverse manner.
rastering
　see aliasing.
rate map
　a map with road segments marked by average speed or posted speed limit along the segments (such as Plate XI). Such maps are used to

create isochrone maps showing locations of equal time separation from a given point such as a retail store or school. Each point in a road network is likely to have a unique isochrone map but the collection of such maps may all be constructed from a single rate map.

ratio

generally, the quotient of two numbers or expressions; also, a measuring scale on which data points are distinguished by size of value, and on which multiplication and division of data point values are well-defined. A scale measuring age is an example of such a scale.

ratioing

analysis of remotely sensed images involving the calculation of the ratio of spectral reflectances in two different bands of the spectrum.

rational numbers

numbers that can be written as the quotient, p/q, of two integers, p and q, with q non-zero. This set includes all fractions, and integers. Between any two rational numbers there is another rational number.

Rayleigh scattering

one form of scattering of a lightwave in which distances between scattering centers are small compared to wavelength.

RBV

see Return Beam Vidicon.

Re-enterable In Cable Splice, RICS

an auxiliary splice closure placed between a main splice and a lateral cable to avoid the problems created by repeated re-entry into the main splice. In this method, all complements that could eventually be committed to a distribution area are connected in the main splice to a stud that ends in the RICS closure. Commitments then are made in the RICS closure without disturbing the main splice closure.

Read-Only, RO

a teleprinter receiver without a transmitter.

Read-Only Memory, ROM

nonvolatile semiconductor storage device with predefined content that cannot be erased.

ready access terminal

differs from a fixed count terminal in that accessibility is available to any of the pairs in the cable.

real number system

the set of numbers that is formed from all the integers, rational numbers, and irrational numbers. A useful characteristic of real numbers is that they form a dense set--between any two real numbers there is always another one. This is of course not true for integers. Thus, in a mapping context, real numbers are useful for situations requiring labelling of continuous phenomena (for example), while integers are useful for labelling discrete phenomena.

real-time system
computer system in which outputs follow inputs with only a small delay.

rearrangement
the process of disconnecting some cable pairs at splices and reconnecting them to other cables. Rearrangement may be required for relief or because of shifts in demand patterns.

reassignable secondary distribution pairs
cable pairs, other than primary or permanent secondary distribution pairs, within the distribution network. They are used for additional business or miscellaneous lines. Because these cable pairs are not cut off beyond their service connection location, they are available for reassignment at other locations in the same distribution cable area.

receiver
equipment that accepts a signal, as from a satellite.

reclassify attributes
change in value of a set of current attributes, often user-specified changes.

record
a logical unit of data in a file.

rectify
to remove geometric distortion, often by aligning raster features or vector coordinate with features in a coordinatized framework. More generally, to superimpose locations on one image or layer with corresponding locations on another image or layer. See Plate II.

Red-Green-Blue, RGB
the additive color system used by color monitors (CRTs).

redirector
software that captures workstation application requests for services like printing and sends them to network devices.

Reduced Instruction Set Computing (or Computer), RISC
a computer architecture characterized by a limited number of hardwired instructions, most of which execute in a single machine cycle and can be strung together quickly to perform complex operations in one processing cycle; opposite of CIS.

redundancy
inclusion of duplicate entries. In a network, redundant connection may offer extra transmission capability or extra security; in a data base, redundant entries may offer no addition to the information content.

reference image
image on CRT used as visual reference.

reflectance

ratio of energy reflected by a body over energy incident with the body. White pin-points in the highly urbanized area on Plate III suggest points with a high level of reflectance.

reflection

radiation that is neither transmitted nor absorbed.

refraction

bending of non-normal incident electromagnetic rays or waves; the rays or waves are usually bent crossing the interface between media of different refractive indices.

refractive index

the ratio of velocity in a vacuum to its velocity in the transmitting medium.

refresh

raster-scan graphic display monitors create an image that is refreshed or redrawn as a series of horizontal scan lines. When the pattern of refreshing is sequential, the monitor supports a non-interlaced pattern of presenting the horizontal scan lines in numerical sequence; in an interlaced pattern, lines 1, 3, 5, 7, ... are refreshed and then the pattern begins over again with lines 2, 4, 6, 8, ..., a strategy which may be disturbing to some individuals.

refresh rate

defined by horizontal and vertical scan.

register

a storage location in hardware logic other than main memory; also, to make fit, as layers of maps superimposed on one another.

register marks

tic marks used to align layers of a map so that they register accurately.

registration

geometric alignment of image data so that superimposed features line up accurately. See Plate II.

regression

statistical technique to examine the linear relationship among two or more variables. A line of best fit is found using least squares, and the extent of relationship is the extent of correlation between the variables.

rehabilitation

the restoration of deteriorated existing plant to an acceptable working condition. Often referred to as "rehab."

relation

see function.

relational database

a database structure in which data is stored in tables; the entries in any table can be uniquely identified by an element (or combination of elements) called the key. One of the values of relational databases is that data need only be stored in one place and then located via the relations between tables. The rules for forming the relationships have been investigated extensively, originally in 1970 by E. J. Codd, who developed a number of "normal forms" in which relations must be in order to form a good database. See references to Codd and to Date. Digital mapping products which have a relational database behind them can support making maps showing a wide range of data at various levels of clustering.

relational operators

examples in a GIS are greater than, maximum, and minimum.

relief

the variation in elevation between the high and low parts of a surface. As is the case in looking a global and local maxima and minima of a surface, here too, relief may be local or global. See Plate IX.

remote communications

communications among terminals or computing sites that are remote from the central computer.

remote file service

distributed file system like Sun Microsystem's NFS that lets workstations use remote resources as if they were local.

Remote Job Entry, RJE

entering of batch processing jobs via an input device that has access to a computer through a data link.

Remote Sensing, RS

analysis and interpretation of images gathered through remote techniques, without having the sensor in direct contact with the objects, such as by photographs from satellites. Such photos often are used to collect data concerning the earth's resources.

Remote Terminal, RT

the part of a subscriber loop carrier system that is placed at a site distant from the central office and the carrier system's central office terminal. The remote terminal and the central office terminal are usually connected by wire pairs that carry multiplexed digital or analog telephone conversations between the two terminals. The RT, located in a cabinet or hut, contains electronic circuits that demultiplex the carrier signals and maintain both the RT and the digital lines.

repeater

a device which boosts the amplitude of a transmission or signal on a carrier, thereby allowing a communication network to be extended in length or configuration for computer networks. It contrasts with a regenerator which reforms and amplifies waves. Functions at the physical layer level of the OSI model.

report

a tabular form for ordering and displaying data from a set of files.

Report Program Generator language, RPG

a high-level language designed specifically for writing programs which produce reports.

Representative Fraction, RF

a way of communicating map scale; if 1 foot on the map represents 60,000 feet on the earth, then the representative fraction is 1/60,000 or 1:60,000. Units must be the same in numerator and denominator; a map on which 1 inch represents on mile has representative fraction of 1:63, 360.

Request For Information, RFI

a formal request for information concerning computer system capabilities, vendor services, or system design.

Request For Proposal, RFP

a formal request to vendors to provide a proposal for services.

Request For Quotation, RFQ

a formal request to vendors to provide a bid for specified services.

Request To Send, RTS

a modem interface signal.

resample

interpolate cell values in a raster and thereby form a raster with larger or smaller cells; used in rectification and registration of images.

residual

statistical term used to capture the idea of the difference between an observed and a calculated result.

resolution

a measure of the accuracy of a graphic display as a measure of the ability of a peripheral to discriminate between values; industry standard categories (expressed in pixels per line, p.p.l.) are: low--up to 256 p.p.l.--grainy texture; high--257 - 999--snapshot photographic quality; very high--1000 - 1600; ultra high--1600 +.

resource class

in a LAN, a collection of ports that offer similar facilities.

response time

elapsed time, from propagation and other delays, between the last character of a message and receipt of the first character of the reply.

Return Beam Vidicon, RBV
    camera tube which offer the highest level of resolution television imagery; used in LANDSAT satellites.
reverse
    interchange of dark and light tones, or opaque and transparent areas, on an image.
revolution
    period of time required by a planet to complete one cycle in its orbit; the period of revolution of the earth is about 365 days.
RF
    see Representative Fraction.
RFI [1]
    see Request For Information.
RFI [2]
    see Radio Frequency Interference.
RFP
    see Request For Proposal.
RFQ
    see Request For Quotation.
RGB
    see Red-Green-Blue.
ring
    a closed loop network topology in a Local Area Network in which stations are connected to each other as appendages to a ring-shaped circuit. For a LAN, ring, bus and star topologies are in use.
ripple
    the procedure invoked to make a cable throw or count change.
RISC
    see Reduced Instruction Set Computing (or Computer).
riser shaft
    an enclosure that houses cable being distributed from one floor to another in a building.
RJ-11
    telephone jack.
RJE
    see Remote Job Entry.
RLL
    see Run-Length Limited.
RO
     see Read-Only.
rodding
    the threading of conduit with a device used to install a pulling line.

ROM

    see Read-Only Memory.

rotary

    set of telephone lines identified by a single symbolic identifier; connection is made to the first available free line.

rotation

    length of time it takes for a planet to make one complete turn on its axis; for Earth that time is about 24 hours.

rotational latency

    time spent waiting for a specific track segment to reach a disk drive read/write head.

route

    a part of the cable network between the wire center and its customers. More generally, a path taken from one location to another. Plate VII shows bus routes.

router

    device for large networks that chooses the best data path for packets going through a LAN; operates on the OSI network layer.

routing node

    routing nodes (routers) are nodes (machines) that can send and receive data from one node to another; routers have two or more circuits; routers regularly receive and maintain information about other nodes in a network.

row

    horizontal list of cells in a raster.

RPG

    see Report Program Generator language.

RS

    see Remote Sensing.

RS-170

    a standard for red/green/blue analogue video in the USA.

RS-232

    see RS-232C

RS-232C

    recommended Standard 232 of the Electronics Industry Association that specifies a 25 pin standard connector for transmission between a computer and peripherals such as printers and plotters; the C is generally dropped from the name of this cabling configuration for asynchronous transmission in computer systems.

RS-422

    Recommended Standard 422 of the Electronics Industry Association concerning cable lengths beyond the RS-232 standard.

RS-423

see RS-422; a companion standard to RS-422, not in wide use.

RS-449

Recommended Standard 449 of the Electronics Industry Association concerning the mechanical characteristics of connectors.

RT

see Remote Terminal.

RTS

see Request-To-Send.

rubber sheeting

a topological process in which map features on a raster are adjusted to fit a pre-determined geometric, digital base; this may be viewed as a transformation of coordinate systems. As an example, fire stations might be planned to be at the intersection of every eighth block of a Manhattan grid which is then rubber-sheeted to fit an actual street pattern that is not exactly a Manhattan grid. This GIS tool is very important in offering flexibility in resource and urban planning. It is basically an application of the idea of homeomorphism, so that extension of GIS theory in this direction might search among the various concepts and theorems dealing with homeomorphisms.

run-length coding

procedure to reduce raster storage needs.

Run-Length Limited, RLL

a method of encoding and compressing digital data to reduce recording and storage requirements; a specific version of this encoding procedure especially popular for disk drives is RLL 2/7.

rural wire

paired insulated wire used for distribution in rural areas.

S s

SAA

see System Application Architecture (IBM).

SAC

see Serving Area Concept.

SAG

see Street Address Guide.

SAI

see Serving Area Interface.

sample

in a raster, a group of cells of a single type.

sampling
   selecting a subset of data points from which one can reconstruct, with fair accuracy, the entire picture.

SAR
   see Synthetic Aperture Radar.

satellite
   see Global Positioning System.

saturation
   a measure of distance a given color is from a gray of equal intensity; third coordinate in HIS scheme to define a display color.

Scalable Processor ARChitecture, SPARC
   Sun Microsystems' defined reduced instruction set computer architecture.

scalar
   a numerical value.

scale
   the ratio of a distance on a photograph or map to its corresponding distance on the ground; the scale of a photograph varies from point to point because of displacements caused by tilt and relief, but is usually taken as f/H where f is the principal distance of the camera and H is the height of the camera above mean ground elevation.

scan line
   swath on the ground recorded by overhead scanning device.

scanner
   a peripheral that automatically converts--digitizes--analog data to digital data, generally in the form of pixels (see raster data). Scanners often use an optical laser to scan a paper map, in a line-by-line fashion, and convert what it "sees" to digital raster data, recording a value of light or dark for each pixel it "looks" at. Some are OCR (Optical Character Recognition) scanners. Also, detector in aircraft or satellite used to remotely scan the earth.

scanning
   see scanner.

scatter diagram
   dot diagram showing positions of locations of coordinate pairs in a Cartesian coordinate system.

scattering
   diffusion of incident radiation in various directions caused by suspended particles.

scattering center
   place in the microstructure of a transmission medium at which light waves are scattered.

scattering loss

power loss in an electromagnetic wave from reflections and deflections of the waves from various causes.

scene generation

simulation of a scene from map data, often displayed as an oblique view. See Plate IX.

schema [1]

characteristics of files of nongraphic attributes such as names of data elements and other information required by the software to process attribute data.

schema [2]

the outline of the overall logical structure of the Data Management Retrieval System data base.

scratch file

temporary file.

screen, contact

camera screen used to convert continuous tone to discrete dot (or other) patterns.

screen, halftone

network of intersecting sets of parallel lines (usually intersecting at a 90 degree angle) used to convert a continuous tone to a discrete pattern of variable-sized dots.

screen, tint

discrete pattern, such as closely spaced narrow lines or small dots, used to create the illusion of a solid color.

screen angle

angle at which CRT is placed; desirable angles avoid moiré effects.

screen copy device

used to print a small-format hard copy of a graphics display screen.

scrolling

new lines that are added following the last line on a display screen; when the screen is full, all lines are moved up one line to make room for a new line at the bottom.

SCSI

see Small Computer Systems Interface.

SDLC

see Synchronous Data Link Control.

sea level

a water boundary that is the "zero" elevation.

seam

line of overlap between combined raster objects. Seams are evident in Plate II.

search

with large data bases, an important issue is to structure some efficient way to search them for desired information, lest storage capabilities outstrip the capability to find large volumes of data. A common search procedure is a Boolean search; in it, the computer matches every instance of a user-assigned word (for example) to every document title in a data base. This strategy produces results that are much quicker than non-computerized searches. However, it can also introduce a great many undesired alternatives and miss others that are not exact matches but might be very closely related. One solution to this dilemma is to develop a very good sense of how to choose key words. Another is to develop alternate search techniques that are context-sensitive and reward the user with titles that are "close" in meaning to the target word--this sort of idea employs notions of point set topology (see Miller 1993, describing work of Stephen Gallant; also, Arlinghaus, S., 1986).

search by attribute

capability to search a database of a GIS for objects with specified attributes.

search region

area to be searched for data points, such as a circle, used to estimate the value of some interior point, such as the circle center. This single value might then be used to classify the local region.

section

basic unit of a square of land one mile on a side in the U.S. Survey of Public Lands, completed in the 1800s. Ownership descriptions of lands are given in terms of sections and fractions (expressed in powers of 2--halves, quarters, ...) thereof. See parcel.

SEED

see Self-Electro-optic-Effect Device.

seed file

a model file used to construct other files with the same characteristics.

segment

a finite portion of a straight line, geometrically. It is in the geometric regard that this word is commonly used. One can view a geometric segment as a restriction of an infinite line to a bounded subset of the larger entity. From this vantage point, in a mapping context, a segment refers to a user-selected, restricted, portion of a raster. Thus, one might select sidewalks, as a segment, from a map of a city neighborhood; a segment may have holes in it, as does a typical segment of sidewalks (from street crossings).

selection query
in a database query, a query formed from multiple comparison expressions; the element is included if the entire query is non-zero.

selective sampling
data sampling technique.

selector channel
an input/output channel designed to operate with only one input/output device, as opposed to multiple input/output devices, at a time.

selector lightpen
special feature that can be attached to a display station; used to point at a portion of the image on the CRT which is then identified for subsequent processing.

Self-Electro-optic-Effect Device, SEED
used in all-optical regenerators.

semivariogram
figure displaying the relation between variance and separation distance.

sensor
equipment to detect energy and process it into a form designed to obtain information about the environment.

sensor platform
physical structure that holds sensor, such as an airplane, satellite, or space station.

SEQUEL
see Structured English QUEry Language.

serial class
in GIS usage, statistical class in which intervals have limits in numerical relation to each other--such as standard deviations.

serial communications
see serial transmission.

Serial Line Internet Protocol, SLIP
a variation of TCP/IP used in X terminals to let them transmit data packets and access several hosts through an RS-232C serial communication line.

serial port
connection between computer and peripherals--bytes are transmitted serially, one following another, through a single transmission line. The difference between serial and parallel port is similar conceptually to the difference between holiday decorative lights that are strung in series (so that when one burns out they must all be tested) on in parallel (only the bad one goes out). It is also conceptually similar to the difference between 1 lane and 6 lane roads.

serial transmission

electronic transmission procedure in which each bit is sent sequentially (serially) over a single channel (either asynchronously or synchronously), rather than simultaneously over multiple channels (as in parallel transmission). Serial lines are often used to join peripherals to computer networks.

server

any networked computer often dedicated to providing file, print, mail, or similar services; a term sometimes applied to a primary file-sharing computer running a centralized network operating system.

Server Message Block, SMB

a distributed file system that lets workstations use remote devices, files, and applications as if they were local; LAN Manager uses SMB at a very low level.

service wire

a small buried wire facility used to connect the customer's premises to a distribution terminal. (If placed aerially, this line is called a drop wire.)

Serving Area Concept, SAC

a technique, introduced in the early 1970s, of administering telecommunications outside plant. It is also referred to as interface design because it features interface units (also called serving area interfaces) between the distribution plant and feeder plant. The Serving Area Concept (SAC) requires that a wire center be divided into serving areas of 200 to 600 housing units, each served from a serving area interface. The area served from the interface becomes the distribution area. In SAC, each living unit is usually assigned at least two pairs of wire in the distribution plant, and the feeder plant provides an average of one and one-half pairs per living unit back to the wire center. SAC reduces plant-operating costs and improves utilization of feeder pairs.

Serving Area Interface, SAI

a device for interconnecting feeder cable pairs to distribution cable pairs of a serving area. Introduced into new or existing plant, interfaces are established under engineering work orders. Serving area connectors are represented by the symbol X on outside plant records and construction work prints. Also see interface.

session layer

fifth level in OSI reference model; used for administrative tasks like security.

shaded relief map

map with areas shaded according to varying levels of elevation; shading is done assuming a certain perspective with an assumed light source fixed in a location (to determine shadow position). In Plate IX, the dark sides of the ridges are shaded assuming a light source in the upper left hand corner of the diagram.

shape

spatial form.

shared access

access method that permits multiple stations to use the same, shared, electronic transmission medium.

shared visual space

capacity of a system for users to interact with a common graphics display area. A change made by a single user is seen simultaneously by all users.

sheath

the protective outer covering of a cable core.

shell [1]

a system command language interpreter that implements a user interface between the operating system and the user.

shell [2]

an outer layer of a program that provides the user interface, or the user's way of commanding the computer; also kernel.

shielding

a protective covering that protects a circuit from unwanted electromagnetic or radio frequency interference.

shift

see FIGures Shift or LeTteRs Shift.

short-haul modem

see line driver and local dataset.

sidelooking radar

remote sensor effective in obtaining images of terrain of relatively large geographic areas.

SIF

see Standard Interchange Format.

signal-to-noise ratio, S/N, SNR

index measuring (in deciBels) strength of desired signal relative to undesired noise.

signature

spatial and other related characteristics that, as a group, identify a feature by remote sensing.

signon character

character used to determine the data rate.

similarity matrix

a square matrix of dimension $1/2*N*(N-1)$ where N is the number of entries in the associated data base; the matrix dimension is the sum of the first N-1 integers.

Simple Network Management Protocol, SNMP
an application-level protocol that allows logically remote users to inspect and alter network-management variables.

simplex
basic unit--cell--on which the combinatorial approach to topology is structured; plural is simplices; adjectival form is simplicial, as in simplicial complex, a cellular topological structure.

simplex method
see linear programming.

simplex transmission
transmission in one direction only.

simulation
mathematical models for studying possible outcomes in a given situation; in a GIS, this strategy is used in digital models of landscapes.

Single Mode Fiber, SMF
a fiber wave guide through which only one optical mode will propagate; permits signal transmissions at extremely high bandwidths. A single mode optical fiber optimized for 1300 nm transmission.

Single Mode Fiber/Dispersion Shifted, SMF/DS
a single mode optical fiber optimized for 1550 nm transmission.

single mode launching
insertion of a single propagation mode into a fiber optic waveguide, with the method of insertion controlled in a variety of ways.

Single Message-unit Rate Timing, SMRT
U.S.A. telephone tariff.

Single Side Band, SSB
a type of radio signal transmission based on amplitude modulation.

size of cable
number of pairs of wires inside the sheath--2 wires required to serve 1 phone=1 pair. Sizes range from 2 to 3600 pairs.

skeletonizing
storing outline of a geographic area by contracting it topologically to a line.

skew
suggests a certain lack of symmetry; lines in three dimensional space, such as a northwest to southeast diagonal on the top of a cube and a northeast to southwest diagonal on the bottom of a cube, that do not intersect and yet are not parallel are said to be skew lines.

slant range
in a radar image, the distance along a line joining the antenna to the target object.

slave station
station controlled by master station (in point-to-point circuits).

SLC
see Subscriber Loop Carrier.
SLIP
see Serial Line Internet Protocol.
sliver
a scanning error resulting in a gap between two lines.
sliver polygon removal
in polygon overlays, slivers may result; removal of slivers is generally desirable.
slope
measure of line steepness; in Cartesian coordinates, the change in y over the change in x; the change in the vertical component over the change in the horizontal component. Slope, m, of a curve, at a point (x_1, y_1) is measured as the slope of the line tangent to that curve at the point of interest; this is found by finding the derivative of the equation for the curve and evaluating it at the point of interest. This value is the slope of the tangent line. The equation for the tangent line may then be found using this slope and the coordinates of the point of interest in the point-slope form for the equation of a straight line: $y - y_1 = m*(x - x_1)$.

slow scan video
equipment that transmits or receives still video pictures across a narrowband telecommunications channel.
SMA
see Standard Metropolitan Area.
Small Computer Systems Interface, SCSI
a bus-based high level computer-peripheral interface.
small scale
a map of scale with a relatively small representative fraction; 1:10,000 is smaller than 1:5000. In the corresponding maps of the same physical dimensions, the area covered in the small scale map is more extensive than it is in the larger scale map; however, the level of detail shown in the small scale map is less than it is in the large scale map.
smart line following
process in which user clicks on lines to convert raster line elements to vector line elements. Uses template matching to follow line and "jump" gaps; operator intervention needed at intersections to restart follower/tracker. Examples--CadCorp-Tracer; LaserScan--UTRAIC.
smart terminal
able to perform its own computations and communicate with host.
SMB
see Server Message Block.
SMD
see Storage Module Device.

SMF
see Single Mode Fiber.

SMF/DS
see Single Mode Fiber/Dispersion Shifted.

smoothing
technique for making a curve appear smoother by removing local variation from lines or data sequences, using moving averages or some other means of filtering out high frequency data components.

SMRT
see Single Message-unit Rate Timing.

SMSA
Standard Metropolitan Statistical Area; see SMA.

S/N
see Signal-to-Noise Ratio.

SNA
see Systems Network Architecture.

snap distance; snap circle
distance within which one feature is made to coincide with another during digitizing; thus, end nodes on line within the snap distance of another node will be moved together to form a continuous path. This function enables one to create closed curves, for example, even when the digitizing is difficult. The user can set the snap distance (radius of snap circle).

snapping
process of moving a feature to coincide with coordinates of another feature.

Snell's Law
a law of refraction which describes the relation between the incidence angle and the refraction angle of a ray of light passing from a given transmission medium to a denser medium. In passing from the less to the more dense medium, the path of the light is bent away from the normal.

SNMP
see Simple Network Management Protocol.

SNR
see Signal-to-Noise Ratio.

soft copy
temporary copy of an image, such as that which appears on a display screen.

soft font
see downloadable font.

software

general name for the set of programs, documents, languages, and procedures involved in the operation and maintenance of a data processing system; these are stored on some sort of medium and can be loaded from there into RAM for execution.

solid color

in a computer, a color formed from pixels all of which are the same color.

solid modeling

creation of a three dimensional surface from a database structure (usually vector). Might be used to create a three-dimensional model, viewed on a two dimensional medium such as a CRT or paper, of a part of the earth's surface. See Plate IX.

solstice

see circle of illumination.

SONET

see Synchronous Optical NETwork.

source code/language

structured program written in a third or fourth generation computing programming language with commands close to human language. FORTRAN, COBOL, BASIC, and C are examples of source code. This sort of code must be translated to machine code before it can be executed by the computer processor.

source IPID

the IPID number of the item of plant that is the source from which another cable derives its count; a mechanism for establishing linear feature connectivity using data base attribute values rather than x,y-coordinates.

space

blank character, ASCII or EBCDIC; also, a binary 0.

spaghetti

a digitizing process in which the operator creates map-spaghetti, a batch process linking points, lines, and polygons in a topologically structured data base that does not require the operator to directly relate line segments to the polygons that use these segments within their boundaries.

SPARC

see Scalable Processor ARChitecture.

spatial

an adjective referring to "space;" as contrasted with "temporal," an adjective referring to time. Often geography and history are concerned with many of the same issues; whether one adopts a spatial or a temporal perspective may guide the direction of useful related matter.

Geography studies spatial relations among phenomena distributed usually, but not necessarily, in two or three dimensional space.

spatial autocovariance

notion that locations close together in space are more likely to have closer values (of whatever is being measured) than are locations farther apart. This idea is certainly not universal, as electronic communications via Bitnet/Internet illustrate; I may have more in common with an individual in the Netherlands than I do with my neighbor in Ann Arbor.

spatial correlation

statistical measure of spatial association between geographic phenomena.

spatial data

data with the attribute of location.

spatial filtering

operations that enhance or block spatial frequencies within an image; these operations include low pass and high pass filtering.

spatial query

use of user-designed or pre-defined shapes to extract cartographic data from an image.

spatial resolution

capability of an image system, such as a satellite, to separate closely spaced objects on the ground.

spatial statistics

statistical techniques that take account of location: for example, the idea that locations that are close to each other are likely to experience greater interaction.

spatial variable

variable with values that vary across an area.

SPC; SPCS

see State Plane Coordinate system.

special characters

characters not on a computer keyboard.

special circuits

cable pairs used by a customer for other than talking lines. Examples: fire alarms, burglar alarms, radio lines, and TV.

spectral band

continuous wavelength range in the spectrum of electromagnetic energy.

spectral bandwidth

the spectral bandwidth for single peak devices is the difference between the wavelengths of adjacent bands at which the radiant intensity is 50% (unless otherwise stated) of the maximum value.

spectral response

the responsivity of a photodetector to various wavelengths of incident electromagnetic radiation, usually indicated as the electrical current or power output as a function of incident power input.

spectral signature

measure of remotely sensed characteristics of a target object at a number of different wavelengths.

spectrometer

radiometer with a prism or other element to disperse light so that properties of incident radiation can be determined as a function or radiation wavelength.

spectrophotometer

equipment to measure relative proportions of wavelengths of electromagnetic radiation in a color.

spectrum efficiency

number of users/services that can be supported simultaneously by a limited radio frequency bandwidth in a defined geographic territory.

speech plus

technique for combining voice and data on the same line.

speed

see data rate.

speed dialing

dialing method using short sequences of numbers to represent complete telephone numbers. Often available on home telephones, as well as in professional environments.

speed of light

2.998×10^8 meters per second.

sphere

three-dimensional object often used to represent the surface of the earth. Because that is the case, it is often the domain of the space used to create flat maps as a projection (mathematical transformation) of the sphere in the plane. For many purposes the sphere is an adequate average surface. For global issues, the fact that mountains and valleys vanish is often not important; however, the curvature may well be significant. For local issues, the reverse is true--the loss of topographic variation (under and above the water) might well be important while the earth's curvature might well not be important (also see State Plane Coordinate system). As in any form of approximate representation, there are trade-offs; the sphere is easy to work with, particularly in theoretical matters.

spike

scanning error in a GIS resulting in a line overshooting its target.

splice

the point at which two or more cables are joined. The physical connection of pairs of wires in a cable.

splicing loss

an insertion loss caused by a splice in an optical fiber; these losses range from a fraction of a decibel to several decibels.

spline

derived from a cartographer's spline, a wooden instrument for connecting separated points not lying on an easy-to-draw curve, this term generally refers to any of a variety of mathematical techniques for interpolating values between a given, finite set of values, in order to represent spatial variation smoothly.

spooling

the technique of using a high-speed storage device to buffer data passing between low-speed I/O devices and high-speed main memory; this procedure enables the user to print in the background while continuing to work in the foreground. The low-speed devices, which can be either the ultimate sources of the ultimate destinations of buffered I/O data, are called spooled devices; the high-speed mass storage devices are called intermediate devices.

SPOT

Système Probatoire d'Observation de la Terre. French satellite, operating since 1986, sending back images with 10 (monochromatic) and 20 (multispectral) meter resolution. See Plate III.

spot elevation

an elevation point established by survey or interpolation; also termed a spot height.

spot height

an elevation point established by survey or interpolation.

spread spectrum

method of signal transmission over a broad range of frequencies. A compatible receiver is used to reassemble the signal. This provides reduced interference and increases the number of simultaneous users with given radio frequency bands.

spur

in a GIS, a piece of a line, not connected to an end-node, jutting out beyond a three-way intersection at a node. To make a clean map, these should be removed.

SQL

see Structured Query Language.

SRDM

see SubRate Data Multiplexer.

SSB

see Single Side Band.

SSI

Small-Scale Integration. See Large-Scale Integration.

standard

agreed-upon model against which others are to be tested. Some standards are the result of consensus of users while others are established by organizations and governments.

standard deviation

in a statistical frequency distribution, measure of dispersion from the mean.

Standard Digital Data Exchange Format

spatial data exchange format adopted by NOAA.

standard error

measures the extent to which a mean calculated from a data set is in fact representative of that data set.

Standard Interchange Format, SIF

spatial data exchange format. A standard or neutral format used to move graphics files between computer systems.

Standard Linear Format, SLF

spatial data exchange format; DMA.

Standard Metropolitan Area, SMA

a U. S. Bureau of the Census term describing an urban area with these characteristics: 1) with a population of 50,000 or more, 2) the county in which it is located, and 3) some adjacent counties; known as Standard Metropolitan Statistical Areas prior to 1980.

star

a LAN network in which the central control point is connected directly to each individual station; graph theoretic structure--on n+1 nodes, one node is joined to each of the n other nodes and none of the n other nodes is joined to each other. Plate VIII shows graph-theoretic stars (spiders) in one layer; this structural model is often used to display journey-to-work patterns in which overall impression, rather than clear views of single edges, is important. A boundary may easily be drawn to separate the stars and assign each to a separate compartment of space: an optimal spatial arrangement. When stars intersect, this strategy is not possible. Thus, any map of in which stars intersect depicts a non-optimal arrangement. In Plate VIII the arrangement of stars is close to optimal; there are only a few intersections.

Starian

an AT&T-developed CSMA network system used on twisted-pair telephone wire.

start bit

in an asynchronous transmission, the first bit, used to indicate the beginning of a character.

State Plane Coordinate system, SPC

National Geodetic Survey Cartesian map projections for U.S.A. Zones in states with a dominant north-south orientation are mapped on a Transverse Mercator projection; those with a dominant east-west orientation are mapped on a Lambert Conformal projection. (The Alaska panhandle is mapped on an oblique Mercator.) The extent of the mapped zones is 158 miles in order to limit the accumulated distortion, that is a result of the earth's curvature (also see sphere). Common in GIS technology.

station

a unit on a network. Also, an installed telephone or other wired communications instrument on a customer's premises. The station has been given a telephone number and is the instrument through which service is furnished to the customer.

STC Cross-Connect

see DSX-0A.

stenciling

way to mask unwanted data on an image.

step-index fiber

a fiber in which the core is of a uniform refractive index, and there is a sharp decrease in the index of refraction at the cladding. Same as radially stratified fiber.

step-index profile

a situation in which the refractive index of a material in a transmission medium changes suddenly, as in a sudden change at a fiber optic core-cladding interface.

stepwise linear classification

see maximum likelihood classification.

stereo compilation

use of photogrammetric equipment to produce a map as a compilation of aerial photos and geodetic control points.

stereo elevation

use of stereo pairs from remotely sensed images to create a three-dimensional elevation surface.

stereoplotter

photogrammetric mapping device used to digitize selected features visible in aerial photographs.

stop bit

in an asynchronous transmission, the last bit indicating the end of a character.

storage

parts of computer used to hold data and programs on a relatively permanent basis (user can delete them).

Storage Module Device, SMD

a device level computer-peripheral interface.

straight splice

the connecting of two cable ends.

strand

a cable used to support aerial outside plant.equipment.

Street Address Guide, SAG

a listing (tape, hardcopy, or map) of the streets and existing addresses in an area (wire center).

stretch

image processing technique to make an actual frequency distribution fit an hypothetical distribution.

string

consecutive sequence of characters; a boundary string is a sequence of arcs.

structural models

graph theory is the most applicable branch of combinatorial mathematics today, with relevance as useful genuine mathematical models not only to virtually all areas of computer science, but also to the social sciences (anthropology, economics, geography, psychology, ...), the physical and life sciences (biology, chemistry, physics, ...), the engineering areas (civil, electrical, mechanical, ...), and the humanities (linguistics, literature, philosophy, ...). The reason for the pervasiveness of this phenomenon is that graphs capture precisely the structure of relationships between entities, i.e., networks of all kinds. The phrase, structural models, refers to those mathematical models which use graph theory. (Definition due to Frank Harary.) On Plate VIII, and on the Cover Plate, graph theoretic stars are used as structural models of the assignment of students to existing schools.

Structured English QUEry Language, SEQUEL

original version of SQL, a high level database language. Developed by IBM for use with the DB2 DBMS.

Structured Query Language, SQL

an ANSI standard language used to implement and obtain information from relational data bases. It was originally developed for IBM as SEQUEL. Most relational databases use some dialect of SQL. SQL can be used both as an interactive query language (interactive SQL) and as a database programming language (embedded SQL).

stub

a short length of cable spliced at a point on a main cable where branch feeder, distribution, or other main cable is or is expected to be connected. Also, a short length of cable that has been factory-connected to a piece of equipment or apparatus.

stylus

cursor used in a digitizer; pen-point.

subfeeder cable

a cable containing 300 to 900 pairs that connects feeder cables to distribution cables.

subrate

in DDS, a data speed of 9.6 kb/s, 4.8 kb/s, or 2.4 kb/s.

SubRate Data Multiplexor, SRDM

a unit that combines a number of data streams at or below some basic rate (2.4, 4.8, 9.6 kb/s) into a single 64 kb/s time division multiplexed signal.

subroutine

a subset of a main program, that is itself a program, and that can be called from the main program.

Subscriber Loop Carrier, SLC

the name for a group of outside plant systems that allows a larger number of customers to be served by fewer wire pairs. The SLC systems electronically combine two or more customer conversations on one or more pair of wires. This allows the telephone company to give each customer a full-time connection to the central office without having to supply each with a separate wire pair. See pair-gain systems.

subtractive color

see color.

Suits-Wagner classification

similar to boxcar classification; see boxcar classification.

Super Video Graphics Array, SVGA

video display with resolution of 1024 by 768.

superimpose

see overlay.

supervised classification

user supervised feature classification in this form of multi-spectral image interpretation.

supplemental contour

a contour used to enhance low elevation areas thereby depicting greater detail between intermediate contours which are spaced at greater distances from each other.

suppress

in a GIS, exclusion of objects by attribute type or object class.

surface fitting

any of various quantitative methods, such as piecewise triangulation, used to interpolate values on surfaces (often by least squares), such as elevation surfaces.

surface mapping

map of three-dimensional information, created using various techniques, such as shading, relief displays, or others. See Plate IX.

surface-emitting LED

a light-emitting diode whose spectral output emanates from the surface of the layers.

surveying

science the involves the precise determination of areas, positions of lines and points, and the topographic variation on the earth's surface. It also involves the accurate representation of the observed field evidence, from the "earth laboratory," on paper or similar media suitable for disseminating and archiving results.

SVGA

see Super Video Graphics Array.

SVHS video

Super VHS video; recent standard for video cameras and recorders.

swapping

the method for sharing memory resources among several processes by writing an entire working set to secondary storage (swap out) and reading another working set into memory (swap in).

swath width

area on both sides of a platform surveyed by remote sensing.

switch

device used to direct packets, generally at a node of the backbone of the network.

switched line

communications link, such as a public telephone line.

switching matrix

electronic form of a cross-bar switch.

SYMAP

see SYnagraphic MAPping program.

symbol

graphic element, such as a diagram, design, letter, character, or abbreviation placed on maps, charts, and others, which by convention, usage, or reference to a legend is understood to stand for or represent a specific characteristic or feature. GISs can display a wide variety of symbols.

symbol overlay

see overlay and overlay analysis; Plate IV.

symbolic name
a identifier for a group of stations (access group) or computer ports (resource class).

SYN
control character to maintain synchronization.

SYnagraphic MAPping program
grid-cell mapping program developed by H. T. Fisher of Harvard.

synchronization
procedure of making two pieces of equipment be in step, relative to time, with each other.

synchronous
having a constant time interval between successive bits or characters; using no redundant information, such as start and stop bits, to identify the beginning and the ending of the unit--a fixed rate method of transmission, often used in mainframe environments and in high-speed LANs.

Synchronous Data Link Control, SDLC
a bit-oriented communications protocol for synchronous data transmission. Developed by IBM--part of its Systems Network Architecture.

Synchronous Optical NETwork, SONET
a standard interface for high-speed, large bandwidth, fiber optic transmissions.

syntax
rules controlling use of statements in a computer language; as in a natural language. Various other linguistic terms concerned with natural language (such as semantic, for example) may be used in connection with computer language.

Synthetic Aperture Radar, SAR
compound sensing system in which one can scan a signal through space without moving antenna parts; a special purpose remote sensing imagery collection device.

synthetic resolution
apparent increase in spatial resolution of remotely-sensed images. This increase might come from any of a variety of techniques applied to the images, including the combining of images from sensors of varying resolution.

System Application Architecture, SAA
specifications for programmers to create a similar look and feel for applications on mainframes, minicomputers, and PCs.

Systems Network Architecture, SNA
an IBM inter-computer communications protocol. Describes the logical structure, as a graph-theoretic tree, formats, operational

sequences, and protocols for transmitting information between software and hardware devices.

T t

T-1

a 24 channel time division carrier system operating at 1.544 Mbps in the United States and Japan and 2.048 Mbps in countries using CCITT standards; uses two cable pairs for two directions of transmission; requires regenerative pulse repeaters at approximately 6000 feet intervals on 22 gauge high capacity cable.

T-Carrier Administration Center, TCAC

center with responsibility for the maintenance and restoration of T-carrier facilities on an automated basis.

table coordinates

tolerance built into digitizing tables, plotters, and similar equipment.

tablet

small digitizer, often at a graphics workstation.

tabular data

data sets that describe digital entities that are not graphic in nature; these might include data involving text or attributes of a map. For example, total population figures are not graphic, but population density figures are as they represent a population value per unit of area.

tag

descriptive element in a database, such as a current data to indicate when changes were last made.

Tagged Image File Format, TIFF

standard color image file format used to transfer images between different software.

tagging [1]

assigning labels to polygons.

tagging [2]

establishing a locus for graphics for the subsequent association of attributes.

tail circuit

a leased line serving as a feeder to a network node.

tap

in a LAN, a cable connection to the main transmission medium.

tap splice

a bridge splice where a new terminal or cable is to be spliced into a through cable.

tape drive

peripheral mass storage equipment. Often reel-to-reel or cartridge tape drives are used as back-up facilities; cartridges may be stacked in juke box format.

taper

a change of pair size at a cable splice or terminal.

taper codes

Bell System standard codes used for both manual and mechanized cable counting and fill-monitoring plans. They are used in Loop Cable Record Inventory System (LCRIS) and also become a part of permanent outside plant records. The rules for assigning taper codes are given in BSP Section 936-312-111 Taper Codes.

taper point

on a feeder route, a point at which the number of cable pairs changes. From the central office to the end of a route, there will be some taper points where the number of cables (and number of pairs) decreases or tapers to a lower value as cables with finer-gauge come to an end.

TARGA file

see Truevision Advanced Raster Graphics Adapter file.

tariff

schedule of rates for services, equipment, and facilities.

tax map

a map depicting the boundaries of each property in a defined area. Tax maps are normally obtained from the county or local tax department. See cadastre. These maps are similar in visual character to Plate V.

TB

see TeraByte.

TCAC

see T-Carrier Administration Center.

TCP/IP

see Transmission Control Protocol/Internet Protocol.

TDM

see Time Division Multiplexing (Multiplexor).

TDMA

see Time Division Multiple Access.

Technical and Office Protocol, TOP

an inter-computer network and communications protocol; a version of MAP adapted to office automation needs.

tee coupler

a coupler for optical fibers which is T-shaped and permits part of the signal power to be reflected from one side of the surface out of the fiber at a right angle in one direction, as it also allows an input signal from the other side of the fiber to be reflected from the other side of the surface.

TELCO

U.S. variant for TELephone COmpany.

telecommunications [1]

the use of wire, radio, optical or other electromagnetic channel to transmit or receive signals for voice, video and data communications; communications over distance, between geographically separate computers (or other electronic equipment), using electrical means.

telecommunications [2]

according to the FCC uniform system of accounts, section 32.9000, "Telecommunications means any transmission, emission, or reception of signs, signals, writing, images or sounds, or intelligence of any nature by wire, radio, visual, or other electromagnetic system. This encompasses the aggregate of several modes of conveying information, signals, or messages over a distance; included in the telecommunications industry is the transmitting, receiving, or exchanging of information among multiple locations. The minimum elements required for the telecommunications process to occur are a message source, a transmission medium, and a receiver."

teleconferencing

two-way electronic communications between two or more groups, or more than two individuals in separate locations.

telemetry

transmission of coded analog data from a remote site.

Telephone Industry Association, TIA

a trade group formed in 1988 by the merger of USTSA and the Information and Telecommunications Technology Group of the Electronics Industry Association.

Telephone Influence Factor, TIF

Telephone Influence Factor of a voltage or current wave in an electric supply circuit is the ratio of the square root of the sum of the squares of the weighted rms values of all the sine wave components (including in alternating waves both fundamental and harmonics) to the root mean square value (unweighted) of the entire wave.

TELeprinter EXchange service, TELEX

network of teleprinters, using Baudot code, joined in an international network using public switched lines.

teleprocessing

same as data communications.

TeletypeWriter eXchange service, TWX

North American network of teleprinters.

TELEX
see TELeprinter EXchange service.

template
a collection of commonly used map features, such as north arrows, neatlines, and so forth serve as a base for common maps; template is a general term used to suggest some sort of standard organizational scheme within which to build. Compare to seed file.

temporary storage
a mass intermediate storage device (see buffer).

tera
prefix for one trillion (10 to the power of 12).

TeraByte, TB
computer unit, 10^{12} bytes.

terminal [1]
a peripheral communication device using a CRT and keyboard, capable of sending or receiving data over a data communications channel.

terminal [2]
in loop plant, hardware specifically designed to facilitate connection and removal of either drop or service wire to and from cable pairs at a particular location. Connections are made on devices that provide secure fastening, such as binding posts.

terminal server
equipment that allows one or more terminals to connect to an Ethernet.

tesselation
geometric process of partitioning a bounded or unbounded area of the plane into tiles that do not overlap and that cover the entire region considered. In Plate VI, a square tesselation is evident.

TeX
powerful (non WSYWIG) mathematical typesetting program available in a variety of formats for mainframes, PCs, and so forth. Copyrights held by Donald Knuth of Stanford University and the American Mathematical Society.

text
string of characters such as this phrase; displayed on screen or using a printer or similar device.

text editor
program that permits input of text as well as interactive editing of text; also, word processor.

Thematic Mapper, TM
multispectral satellite sensing scanner, designed to monitor Earth resources, used on Landsat platform that returns high resolution images.

thematic mapping

use of a geographic variable (theme) to show the spatial distribution of that variable by shading map features according to values of variable classes. Thus, a map in which all countries of the world with over 50 million population are shaded black; all with equal to or under 50 million are shade gray; and, all for which there is no data are left white, is an example of thematic mapping in which the theme is total population.

theme

general topic of a map that displays the spatial variation of a single phenomenon, such as paved area of a city.

theoretical splice

not a physical splice. Used on a work print to show a change in field code, change from underground to buried cable, and so forth.

theory

the set of general, abstract principles that underlie a body of knowledge. Thus, one should be careful to distinguish "theory" from an hypothesis, idea, educated guess, speculation, or hunch.

thermal band

generally, a way to partition the electromagnetic spectrum, especially for infrared wavelengths.

thermal infrared

general term for the set of infrared wavelengths in the electromagnetic spectrum; this set covers the wavelengths of radiation emitted by the Earth.

thermal transfer plotter

a plotter that uses heat, to melt dots of waxy ink, to make plots. It often takes numerous passes through the equipment to make complicated maps.

thermocouple

a loop formed from two thermojunctions.

thermojunction

a junction of two dissimilar metals bonded together to form an interface across which a voltage is developed when the junction is heated.

Thiessen polygons

division of areas into tiles by bisecting with orthogonal lines the lines of minimum distance joining sample points. The orthogonals define the Thiessen polygons. Also Dirichlet regions and Voronoi polygons.

thin Ethernet

Ethernet network that uses smaller diameter coaxial cable than normal; sometimes referred to as thinline or thin-wire Ethernet.

thinning

in a raster object, the removal of extra cells from lines to make the lines exactly one cell wide; done to facilitate conversion to vector format. In a vector object, reduction of coordinate pairs determining lines.

One procedure for thinning pixel arrays is as follows:
Given three nodes, A, B, and C, with arcs joining A to B and B to C; label as x the distance from B to a geodesic joining A to C. If x is greater than or equal to some user-defined value n, then B and its incident arcs are retained; otherwise they are deleted.

three-dimensional data

in a geographic context, often the three coordinates are latitude, longitude and elevation, conventionally denoted as an ordered triple on the Greek letters phi, lambda, kappa. Three dimensional objects are represented by their upper surfaces--functions in x and y.

threshold

binary separation to conver gray scale raster data into binary data; with 256 levels of gray, a value of 150 might be selected as the threshold value above which all cells would be coded 1 and coded 0 elsewhere.

throughput

a measure of processing or handling ability as a function of the amount of data accepted as input and processed as output.

TIA

see Telephone Industry Association.

tic

registration or geographic control point.

tic match tolerance

maximum acceptable distance between an existing tic mark and a tic mark being digitized; outside this distance, the map will need to be registered over again.

TIF

see Telephone Influence Factor.

TIFF

see Tagged Image File Format.

TIGER

see Topologically Integrated Geographic Encoding and Referencing.

tile

one element of a tesselation or a mosaic; see tiling.

tiling

spatial segmenting of an image in which the entire image is covered by non-overlapping smaller tiles of image; contrast with cascading on a computer screen. Also, spatial assembly of large images from smaller images, as a mosaic; raster tiles may overlap, although geometric tiles generally do not. In Plate VI, the geometric tiles do not overlap, although the underlying data entries for rivers must do so in order to create a continuous line (red lines cross white tile boundaries).

tilt

the angle at the perspective center between the photograph perpendicular and the plumbline; also the dihedral (space) angle between the plane of the photograph and the horizontal plane.

time districting

process to define contour distances from known sample points and to define associated districts.

Time Division Multiplexing (Multiplexor), TDM

the process of interleaving a number of digital signals into a single digital stream by an orderly assignment of time slots; this arrangement permits several signals to share a single transmission facility.

Time Division Multiple Access, TDMA

access technology often used in U.S. cellular telephone industry; allows a large number of user to gain sequential access to a single radio frequency channel without interference. It does so by allocating unique time slots to each user within each channel.

time sharing

a method of operating a computer system to allow a number of interactive terminals to use the facilities; these terminals are served in sequence at a pace of high enough speed to make it appear as if all terminals are served simultaneously.

time slice

processor time given to a particular application, often in milliseconds.

TIN

see Triangulated Irregular Network.

TM

see Thematic Mapper.

to-node

the last endpoint of an arc to have been digitized.

token

a packet used in explicit access Local Area Networks.

token bus

a bus topology Local Area Network using token passing for explicit access.

token passing

a network data-access method; computers wait to get control of a "token" that lets them send data.

token ring

a LAN ring wiring topology, developed by IBM, that uses token passing for data access; based on the IEE 802.5 standard. Stations must wait for the token to arrive prior to being allowed access to the network to transmit data.

TOP

see Technical and Office Protocol.

topographic features

in a GIS context, often used to refer to map features that contain elevation information as third dimensional z-values (within the Cartesian coordinate context). In a broader sense, these are features on the surface of the earth that would be visible on aerial photographs, as they do posses an elevation component. Mountains, river valleys, and glacial deposits are all examples of topographic features. Some are evident on Plate II (left side) and Plate IX.

topography

the collective set of the features of the surface of the Earth; its configuration, including its relief, the position of its streams, roads, and cities. It embraces elements of hypsography/hypsometry, hydrography, culture and vegetation.

topological coding

a coding system that maintains the characteristics of direction and connectivity between objects, without regard for distance.

Topologically Integrated Geographic Encoding and Referencing, TIGER

U. S. Census Bureau and USGS computer file system; developed in preparation for the 1990 USA decennial census of the population. This computer system has data for all 50 US states, the District of Columbia, Puerto Rico, U.S. Virgin Islands, Guam, American Samoa, and the Northern Marianas. It contains a digital street map of the U.S.A. with data that can be superimposed at a scale of 1:100,000, including address numbers for the 354 largest urban areas. In addition to the population variables, it also contains spatial information involving numerous other features, as well.

topology

a branch of pure (as opposed to applied) mathematics which has evolved, during the twentieth century, to take on a role of considerable prominence, as it examines global and local relations between mathematical objects that are of fundamental importance--such as connectivity, compactness, separation, deformation, and convergence. Transformations with topological spaces permit scale shifts between global and local.

Generally, topology is a branch of mathematics which deals with the entire pattern of relationships between objects in a (topological) space. Graph theory can be thought of as 1-dimensional topology. Abuse of terminology has led some social sciences to refer to the topology of a network when in fact only the pattern of connection is meant. Because much of topology is not interested in distance, but shape, topology has been called rubber-sheet geometry. A standard joke is that a topologist can't tell a doughnut from a coffee cup, since each has one hole, and were they both made of rubber, one could be deformed into the other.

There are a number of different modifiers which may precede this noun, and often the modifier suggests the sorts of mathematical objects being studied; two common modifiers are "combinatorial" and "point set." In the combinatorial approach, relationships among simplexes (in the plane, a simplex is a triangle), and larger complexes, become important; R. H. Atkin used this sort of topology as the foundation of his "Q-analysis." Later in the twentieth century, as set theory came more into vogue, point set topology became important. In the point set approach, the same sort of rigorous logic that underlies set theory was used to develop topology whose mathematical objects are point sets--in it the concept of "nearness" is examined. Accordingly, a topology T in a set X is defined as a collection of subsets of X satisfying the following set of axioms:

1. The null set and the entire set are both members of the collection of subsets.

2. Arbitrary unions of members of T are once again elements of the collection T.

3. Finite intersections of members of T are once again elements of T.

The pair, (X,T) consisting of a set X and a topology T in X is defined as a topological space. Elements of the topological are "points"; members of T are "open sets" of the topology T.

This sort of idea has theoretical GIS implications as to how one might choose coverings of map data and split it apart into layers to optimize ease in representing various shape concepts in many different possible, desired overlays. See Plate VIII. It also seems to hold great promise for searching databases--when some notion of "nearness" (as nearness in meaning) is important (see *Wall Street Journal*, July 27, 1993).

With the definition in hand, a vast array of theorems, corollaries, and other definitions follow. One idea that might, again in the theoretical GIS context, be important is the notion of separation axioms (usually denoted using a "T" for "Trennunsaxiome"--German for separation axiom) There is a carefully defined hierarchy of ways in which point sets might be separated from each other; one of the better known of these is the Hausdorff separation axiom (T2).

There are many useful references in any university library, under "topology" in the mathematics section of the library; it does take considerable time and effort to work through books at this level, because if one does not have the prerequisite for the material, it must be acquired prior to reading the more advanced material. But, careful understanding of the material leads to accuracy in creative application and to avoidance of abuse of terminology and inferior results.

township

see parcel.

TPI

see Tracks Per Inch.

trace

manual tracing of an image, using a light table, which has as its GIS counterpart the interactive tracing of lines in a raster object to create a vector line.

track

the actual path of an aircraft over the surface of the earth.

tracker ball

interactive hand-controlled device for moving a cursor across a display screen.

Tracks Per Inch, TPI

a measure of data storage density.

training

process in which a receiving modem reaches equalization with the transmitting modem. In a GIS, user-defined group of sample cells, as a prototype in an image, known to represent a particular feature.

training site

recognizable area on an image with distinctive properties that will enable observers to recognize similar areas.

transaction processing

real-time data processing, with no editing or sorting, that processes tasks as they occur.

TRANSCEIVER

combination of the services of a TRANSmitter and reCEIVER in a single piece of equipment.

transcoder

device to convert video signal formats from input to output type.

transducer

a device that converts physical parameters such as temperature, pressure, motion, et cetera into an electrical voltage or current; steps down velocity of signal.

transect
cross section of sampling points along a straight line; similar to a traverse.

transfer function
quantitative method to transfer spatial data from one projection to another.

transformation
mathematical mapping from one set to another which is well-defined; may be one-to-one, many-to-one (not one-to-many), onto, or into.

transformation program
a computer applications software used to change digital data from one format to another; the increasing or decreasing of density of elevation posts or altering distribution of elevation posts.

Transformed Vegetation Index, TVI
vegetation index based on satellite images of prescribed spectral bands.

translation
rigid motion in the plane of a geometric object; literally, a sliding across. Many digital objects are not translation invariant; it does matter where the initial cells are located. Translation invariance is desirable because then a structure developed for one set of values can be employed elsewhere.

translation curve
curve to adjust raster cell values to display values.

transmission block
sequence of continuous bytes transmitted as a unit.

Transmission Control Protocol/Internet Protocol, TCP/IP
an inter-computer communications protocol originally developed at DARPA. It functions in the 3rd and 4th layers of the OSI Model.

transmittance
ratio of radiant energy transmitted through a body to radiant energy incident with the surface of the body.

transparency
a positive image upon glass or film, intended to be viewed by transmitted light; also called a diapositive.

transparent color
see-through color; used in GIS in overlays so that spatial detail of both layers are evident for comparative purposes.

transparent patterns
see-through patterns; used in GIS overlays with the same idea as transparent color (see transparent color).

transport layer

fourth layer of OSI reference model; provides error checking and routing of data packets.

transverse Mercator

a conformal cylindrical map projection, being in principle equivalent to the regular Mercator map projection turned (transversed) 90 degrees in azimuth; this projection system includes 60 north-south zones each 16 degrees wide, centered on a central meridian represented by a straight line (see also, Universal Transverse Mercator). Clearly, there is redundant coverage; the central meridians are spaced 6 degrees of longitude apart. This projection is the basis for the zones of the State Plane Coordinate system.

trapezoid

quadrilateral with one pair of unequal parallel opposite sides and the other pair of opposite sides non-parallel. When the non-parallel pair are equal in length, the trapezoid is said to be isosceles.

traverse

see transect.

tree

a connected graph theoretic structure in which there are no cycles, or redundant linkages. Graph theoretic trees often resemble a natural tree in structure. Bus routes, or a substantial portion of them, are often trees--see Plate VII.

trend surface analysis

fit of a polynomial line or surface through data points using least squares that is used as a statistical model of gradual long-range variations.

Triangulated Irregular Network, TIN

strategy for triangulating a surface based on Delaunay triangulation; eliminates redundant altitudes. Used in creating digital elevation models. See Plate I.

triangulation

see analytical triangulation.

triangulation point

a surveyed spot elevation/location.

tristimulus values

sums of amounts of red, green or blue needed to make an accurate match to a target color.

true scale

identifies where map measurements correspond to actual surface distances (on only a portion of any map).

Truevision Advanced Raster Graphics Adapter file, TARGA file

format used to transport color images.

truncate
to shorten, for example, by replacing a corner of a solid by a plane.
tuple
an ordered pair of elements, as in Cartesian coordinates, might be called a 2-tuple because there are two components to the pair. In n dimensional space, the correct mathematical term for a corresponding structure is an n-tuple. The term, n-tuple, has been shortened in the mapping/computer science community. Also, sometimes used for record.
turn
direction of a path (intuitive use of the word, not graph theoretic) at an intersection of edges in a GIS.
turnkey system
a system with components from various vendors; also, a system from a single vendor that is immediately operable.
tutorial
a graphic display on the graphics workstation that allows an operator to interface with a user command by giving data points on the screen.
TVI
see Transformed Vegetation Index.
twisted-pair wire
two thin wires twisted around each other to create cable for linking things like telephones, PBXes, computers, and terminals; it sometimes has extra shielding; its low cost and easy installation make it a very popular media for wiring LANs. The loosely intertwined wires help cancel out any noise.
two dimensional data
areal data in two dimensions, such as latitude and longitude.
two-wire circuit
a pair of wires which, depending on modem type, can be used for a variety of styles of transmission (including 1-way transmission, half-duplex transmission, and full-duplex transmission).
TWX
see TeletypeWriter eXchange service.
Type-A Coax
a serial transmission protocol.

U u

U-guard
a U-shaped metal protective cover placed over a cable (e.g., a cable going up the side of a pole).

UART
   see Universal Asynchronous Receiver Transmitter.
UHF
   see Ultra High Frequency.
UIC
   see User Identification Code.
ULSI
   Ultra Large Scale Integration; see Large Scale Integration.
Ultra High Frequency, UHF
   band from 300 - 3,000 MHZ.
UltraViolet radiation, UV
   radiation of wavelength longer than that of visible light--outside the visible range.
underground cable
   cable placed in manholes and conduit. Maps such as Plate V are often used to visualize locations of cables of this sort.
undershoot
   an edge that does not extend to reach another node.
Uninterruptable Power Supply, UPS
   device(s) which preserve the continuous flow of uniform quality power to a system.
unit value
   the number of data elements per item. This number is a direct reflection of the number of data elements which would be associated with that item of plant.
United States Survey of Public Lands
   see parcel.
Universal Asynchronous Receiver Transmitter, UART
   a device which converts parallel input into serial form or serial input into parallel form.
Universal Transverse Mercator grid, UTM grid
   a military grid system based on the Transverse Mercator map projection. It applies to maps of the surface of the earth extending between 84 degrees north and 80 degrees south latitude--to attempt to cover any of the surface poleward of 84 degrees results in severe distortion because the cylinder of projection has as its axis the diameter joining the north and south poles of the earth.
Universal Transverse Mercator projection, UTM projection
   a map projection system and associated coordinate system in which central meridians are designated every six degrees and the earth is partitioned into sixty north-south zones each 16 degrees wide (see also transverse Mercator).

UNIX

a multi-tasking, multiprocessing computer operating system; derived from MULTICS; implemented entirely in C language; versions currently available include UNIX System V (AT&T), BSD UNIX (Berkeley), Ultrix (DEC), XENIX (Microsoft), A/UX (Apple), and AIX (IBM); scheduled for unification by COS. This was designed to be a portable operating system developed for scientific applications on large processors and on multipurpose computers.

unsupervised classification

automatic multispectral image interpretation process; cells are statistically clustered into collections without the use of training data. The user can interact only after the process is complete.

updating

additions of new points, lines, areas, or other information to a database.

upper memory area

in a DOS environment, the 384 K of memory directly adjacent to the 640 K of conventional memory.

upper memory blocks, UMB

in a DOS environment, parts of the upper memory area.

UPS

see Uninterruptible Power Supply.

USASCII

see ASCII.

user command

a computer program which may be invoked by a board digitizing technician for graphics manipulations; a user command can replace several manual steps with one automatic procedure.

User Identification Code, UIC

defines a user's membership in a particular group; the UIC consists of two numbers, each ranging from 0 to 377: the first is a group number and the second is a member number (DEC).

utility

AM/FM mapping applications for handling information about public utilities; also, terminology for system capabilities.

utility easement

the right of the utility companies to place facilities on a portion of one's land.

utility mapping

a class of mapping applications for managing information about public utilities--telephone, water, sewer, and so forth. See Plate V.

UTM (grid)

see Universal Transverse Mercator grid.

UTM (projection)
see Universal Transverse Mercator projection.
UTP
Unshielded Twisted Pair.
UV
see UltraViolet radiation.

V v

V.42
a CCITT interface recommendation; an error control specification for modems.
value
in a color, the relative darkness (of lower value) or lightness (of higher value) when compared to a standard scale; second designation in the Munsell color system to coordinatize color.
variable
quantity which can assume any value in an underlying set of possibilities.
variable-length record format
a file format in which records are not necessarily the same length.
variance
statistical measure of the dispersion of observed values about the mean of observed values; calculated as the square of the standard deviation.
vault
location in or near a central office where cable originates.
VAXcluster
a VAXcluster configuration is a computer system combining two or more VAX processors and mass storage servers in a loosely coupled manner.
VDT
see Video Display Unit. (also, Video Display Unit).
VDU
see Video Display Unit.
vector
a mathematical entity showing both distance and direction. In a mapping context, this word has come to refer to a data structure for processing and displaying graphic data, represented as strings of coordinates given the true position of features and their boundaries. In contrast to the raster format, in which reference is to pixels or matrix entries, the vector format is capable of stringing together pieces of line, end-to-end, in order to create a curve that appears smoother than does

its raster counterpart formed from little squares (current shape of a pixel). Vector data define complex geometric entities that can be manipulated on the basis of attribute data; they are very useful in this regard.

From a theoretical standpoint, the distinction between vector and raster is parallel to the approach in linear algebra that distinguishes between combining (using rules of algebra) vectors directly, or combining (using rules of algebra) matrices that represent these vectors. Thus, one place to look for theoretical material related to the problem of shifting from vector to raster format is in the literature of modern linear algebra texts that emphasize a transformational approach (rather than the first-half of the twentieth century approach of "matrix algebra").

In a public health context, a vector is an intermediate host that conveys a disease-causing organism from one animal or human to another.

vector data

data encoded as nodes, line strings, and polygons.

Vector Product Format, VPF

primitive elements of a spatial data base consisting of nodes, edges, faces, and text.

vectorize

to convert raster to vector data.

Venn diagram

set-theoretic diagram to partition sets of elements into mutually exclusive and exhaustive classes based on logical relations: three distinct sets A, B, and C, drawn as intersecting circles in a pretzel-shape, partition the space into those elements lying outside all three sets; those in A and B and C, those in A and B and not in C, those in A and C but not in B, those in B and C but not in A, those in A but not in B or C, those in B but not in A or C, and those in C but not in A or B.

Thus there are eight distinct logical classes in a Venn diagram with three sets (there are 4 classes in a diagram with 2 sets). It is not possible to use a Venn diagram with circles representing the sets for 4 or more sets. One generalization that has been used in Boolean algebra, particularly in minimization of switching circuits, is called a Karnaugh map.

version

number to catalogue software; typically a version number is has both an integral and a decimal part. A change in the decimal part indicates a minor upgrade; a change in the integral part indicates a major change.

vertex

see node.

vertical accuracy
   data position relative to control (i. e., source, input, et cetera) in Z direction.

vertical distance
   difference in y-coordinates in a Cartesian coordinatization of the plane. Often used to refer to elevation.

vertical frequency
   related to scan time of a CRT; measure given in Hz.

vertical survey control monuments
   benchmarks that provide control structure on the earth's surface from which to make precise elevation measurements. The strategy for positioning them must be well thought-out in order to insure their visibility on aerial photographs and so that the density of their placement is appropriate to making good measurements at specified accuracy levels.

Very High Frequency, VHF
   band from 30-300 MHZ.

Very Small Aperture Terminal, VSAT
   a type of satellite terminal used to connect several widely spaced offices or facilities.

VGA
   see Video Graphics Array.

VHF
   see Very High Frequency.

Via Net Loss, VNL
   the lowest loss at which an intermediate trunk may be operated in a multi-trunk connection of DDD network without incurring unacceptable echo or singing.

Via Net Loss Factor, VNLF
   a factor used to determine the lowest loss at which a trunk may be operated in a multi-trunk connection of the DDD network from an echo standpoint.

video capture
   freezing and storing of a video image in the computer--in memory or on a disk.

video card
   electronic circuit board to handle video signals in a microcomputer.

video compression
   a coding technique used to reduce the bandwidth required for the transmission of video images by reducing redundant information within or between video frames; also called bandwidth compression, data compression or bit rate reduction.

video digitizing board
   video interface that constructs digital image from frozen video frames.

Video Display Terminal, VDT
   same as CRT. (Also, Video Display Unit.)

video field
   image of about 480 horizontal lines on a standard CRT. The frame (picture) is refreshed about every 1/30th of one second.

video frame
   complete video image; picture.

Video Graphics Array, VGA
   a type of microcomputer graphics display and/or monitor with horizontal scanning frequency ranging up to 31.5 KHz.

video teleconferencing
   two-way electronic voice and video communication between two or more groups, or more than two individuals, at separate locations.

view port
   window used to interactively access part of the underlying map database.

viewable space
   maximum part of content of a display board that can be seen at once; less than or equal to the addressable space.

viewshed
   area visible from a particular location; a closed curve topologically equivalent (homeomorphic) to a circle might be the viewshed of a point; while a sausage-shape--as a union of topological circles--might be the viewshed of a length of road. Barriers of various sorts cause creases and in the edges and holes in the interior of otherwise geometrically-regular viewsheds. See Plate IX.

vignetting
   a gradual reduction in density of parts of a photographic image caused by the stopping of some of the rays entering the lens, often resulting in a darkening of the image at the corners.

virtual circuit
   circuit established only for the length of the call; this circuit may share a single physical circuit with other virtual circuits.

virtual display
   scheme to display as much of an image as will fit on a monitor while storing the remainder of it in a buffer temporarily.

virtual memory (storage)
   a technique for expanding the capability of a computer system by dividing a program and its data into pages, only some of which resides in internal memory at any one time; addressable space that appears as

real memory, from which instructions and data are mapped into real memory locations.

virtual space

videoconference in which each participant is seen on a separate screen or space.

visible wavelengths

range of electromagnetic spectrum which the human eye can detect.

Visual Display Unit, VDU

a terminal.

VLSI

Very Large-Scale Integration; see Large-Scale Integration.

VNL

see Via Net Loss.

VNLF

see Via Net Loss Factor.

voice-data PABX

a single device combining voice PABX and data PABX.

voice-grade channel

line of minimum bandwidth suited for voice frequencies.

voice-switched video

type of videoconference in which voice signals activate cameras to send a picture of a particular person. All participants cannot be seen simultaneously.

volatile

describes a data storage or memory device that loses its content when power is lost.

volume

three-dimensional counterpart of area, characterized in a mapping context by, for example, latitude, longitude, and elevation. Also, in a computer, an individual storage unit such as a floppy disk.

VOlume ELement, VOxEL

the counterpart of the pixel in higher dimensions. See also polygon, pixel, Jordan Curve Theorem.

Voronoi polygons

see Thiessen polygons.

VOxEL

see VOlume ELement.

VPF

see Vector Product Format.

VSAT

see Very Small Aperture Terminal.

W w

WAN

see Wide Area Network.

warping

process that stretches a GIS object.

watershed

drainage basin; area above a selected point, such as the mouth of a river, that drains into that point (also called seed value).   Boundaries of a watershed generally are interpolated along ridges separating drainage basins. See Plate IX.

wavelength

the distance between the nodes of a wave; the ratio of the velocity of the wave to the frequency of the wave--denoted by a lower case Greek gamma.

Wavelength Division Multiplexing, WDM

the multiplexing of lightwaves in a single transmission medium in optical communication systems.   Each of the waves is of a different length.   They are separately modulated prior to insertion into the medium.

WDM

see Wavelength Division Multiplexing.

weeding

see line thinning.

weighted moving average

average value of an attribute for a given point calculated from values at neighboring data points, giving consideration to their distance from, or relative importance to, the given point.

weird polygon

polygon with interior loops; not a Jordan curve.

wetness

measure of how wet a region is, using the combined wetness of vegetation, surface soil, and biomass.

WGS

see World Geodetic System.

What You See Is What You Get, WYSIWYG

this phrase describes a word processor in which the image you see on the screen is what you will see on the printed page--in contrast to such processors as TeX. With WYSIWYG, form and content of display matches output product form and content.

Wide Area Network, WAN
a computer interconnection system providing shared access to, and movement of data, using common carrier-provided remote lines (digital phone lines, microwave) over some widespread area.

wide drains
rivers, glaciers, wadis, mudflats, and other topographic features symbolized by a double line on a map.

wideband
see broadband.

wild card character
a symbol used with many commands in place of all or part of a file specification to refer to several files or values rather than specifying them individually.

WIMP interface
a program interface that uses windows (W), icons (I), a mouse (M), and pull-down menus (P).

window
part of a raster or vector object displayed on a screen; band of electromagnetic spectrum offering certain optimal enhancing and blocking properties depending on the specific sensor; also, terms related to Windows software from Microsoft.

windowing
capability to move features in an underlying GIS database to particular polygons.

wire
conductors (aluminum or copper).   Aerial wire--wire strung on poles. Buried wire--wire placed directly in the ground. Service wire--wire used from a cable to the subscriber's location. Rural wire--paired, insulated wire used for distribution in rural areas. Station wire--wire used inside buildings to connect telephones and apparatus. Open wire--single, non-insulated wire used in rural areas. Multiple wire--individual insulated wires wrapped together. Used in rural areas.

wire center
a geographical area where all locations are served out of the same central office.

Wired Out of Limits, WOL
a facility modification that required a customer to be served from a location other than the location that should provide the service.

wireframe
familiar fish-net-like graphical representation of a three-dimensional object.

wireless

radio-based system that allows the transmission of telephone or data signals through the air without a physical connection (metal wire or fiber optic cord).

WOL

see Wired Out of Limits.

word

a discrete group of adjacent bytes or bits.

word processor

see text editor.

work print

a detailed drawing or print issued to indicate the addition, removal, or rearrangement of outside plant; for example, an estimate, job, or routine order.

working pair

a cable or wire pair that is energized at the central office.

workstation

vague term, often meaning a grouping of equipment that works interactively with a computer to execute some set of specific tasks.

world geodetic system, WGS

(WGS) 1972 - based on the Transverse Mercator projection and referenced to a common (one) datum for all of the earth's surface.

Work Print Generation, WPG

an AM/FM system functionality related to producing of construction plans and bills of materials.

WORM

see Write Once, Read Many.

WPG

see Work Print Generation.

Write Once, Read Many, WORM

a type of laser disc storage system. It is useful in storing large data sets that are primarily archives and are only rarely changed.

WYSIWYG

see What You See Is What You Get.

X x

x,y coordinates

the location of an object on a Cartesian coordinate grid system. Also see, Cartesian coordinates.

X rays

short wavelengths in the electromagnetic spectrum of length about 10 to the (-4) micrometers.

X windows

a system, developed at MIT, that allows applications to be displayed in windows and shared among different workstations and terminals. Available on many UNIX operating systems

X-Conn

crossconnect.

X.25

CCITT recommendation/ISO standard for wide area networks. There are numerous CCITT recommendations concerning interfaces that are prefaced with an X.

X.25 circuit

circuit using the X.25 level 3 protocol (packet level) and providing for communication over the Packet Switching Data Network (PSDN).

X.400

ISO standard for electronic mail.

X/Open

an association of American and European vendors formed in 1984 to promote open systems.

Xerox Network Services, XNS

distributed network file system that lets workstations use remote resources as if they were local.

XGA

see eXtended Graphics Adapter.

XMS memory

Extended memory; applications work with a memory manager.

XNS

see Xerox Network Services.

XOFF, XON

control characters used to control flow.

Y y

yaw

the rotation of an aircraft about its vertical axis; the rotation of a camera about the Z axis.

YMCK

see CMY.

Z z

z-value

third coordinate in a Cartesian triple (to coordinatize three dimensional space) that typically represents a height value.

zenith

point of the celestial sphere directly above the observer; opposite the nadir.

zero

origin of Cartesian coordinate systems, represented as an n-tuple of zeroes.

zero insertion

insertion of a binary 0 in a transmitted data stream in order to separate data from SYN characters; receiver removes the inserted 0.

zero material dispersion wavelength

electromagnetic wavelength at which there will be no material dispersion--at the point in the frequency spectrum where ultraviolet absorptions stops and infrared absorption begins.

zoom

proportional enlargement or reduction of an image by using a scale change, particularly on a CRT.

Zoom Transfer Scope, ZTS

optical mechanical device for viewing an overlay of two images.

ZTS

see Zoom Transfer Scope.

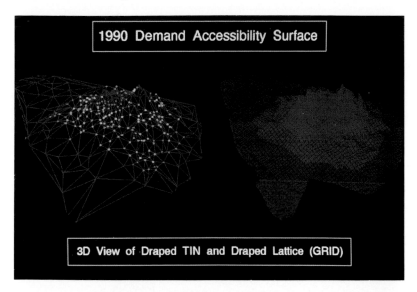

Plate I. Retail business demand/accessibility analysis. Left: 3-D view of draped TIN; right: draped lattice (GRID). Retail analysis using spatial interaction capabilities in Network (ARC/INFO). Points=population, 3D surface=areas of high demand for retail services. Digital image provided courtesy of Environmental Systems Research Institute, Inc.

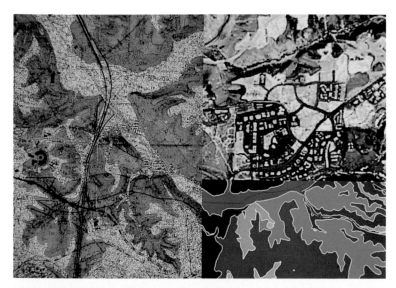

Plate II. Habitat Analysis. Left: scanned geologic map, registered/rectified to geographic coordinates; upper right: integrated satellite image; lower right: digitized vector/polygon, shaded data.. Del Mar, CA. Digital image provided courtesy of Environmental Systems Research Institute, Inc.

Plate III. Multispectral satellite image of San Francisco Bay. Digital image provided courtesy of Environmental Systems Research Institute, Inc. Copyright © 1991-1993 Environmental Systems Research Institute, Inc. All rights reserved.

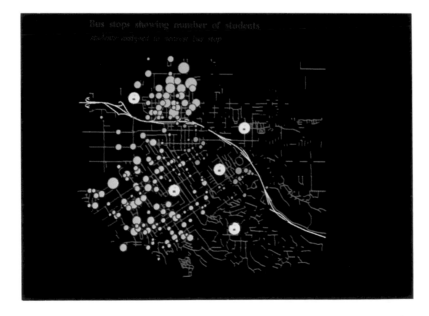

Plate IV. School districting analysis. Five schools and their locations; students allocated to schools based on location of residence. Digital image provided courtesy of Environmental Systems Research Institute, Inc. Copyright © 1991-1993 Environmental Systems Research Institute, Inc. All rights reserved.

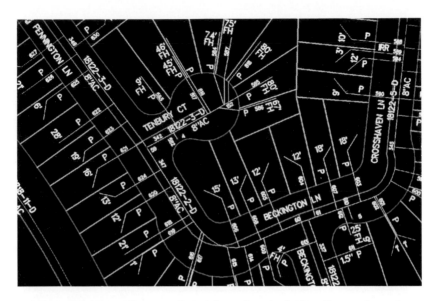

Plate V. Water/gas utilities analysis depicting main and subsidiary lines serving various parcels. Digital image provided courtesy of Environmental Systems Research Institute, Inc. Copyright © 1991-1993 Environmental Systems Research Institute, Inc. All rights reserved.

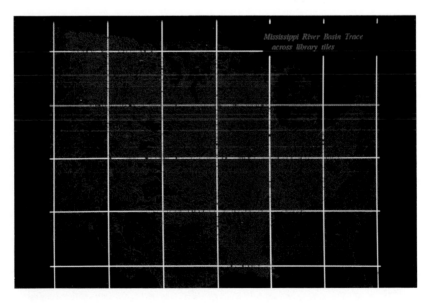

Plate VI. Connectivity analysis. Mississippi River Basin coverage area, library of coverages for river system. Point to area of river, and select all spatial data connected. TRACE capability in ARC/INFO used. Digital image provided courtesy of Environmental Systems Research Institute, Inc. Copyright © 1991-1993 Environmental Systems Research Institute, Inc. All rights reserved.

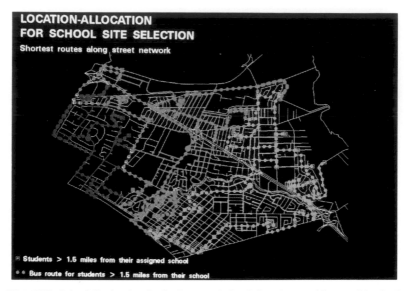

Plate VII. School district planning/busing. Analysis of all students residing outside of 1.5 miles of school sites. Allocation of students to bus routes, assignment of bus routes by residence/school locations. Performed using Network capabilities in ARC/INFO. Digital image provided courtesy of Environmental Systems Research Institute, Inc. Copyright © 1991-1993 Environmental Systems Research Institute, Inc. All rights reserved.

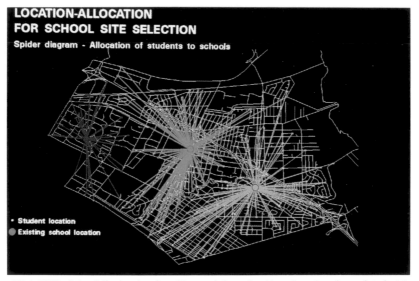

Plate VIII. School district planning. Three existing school locations, locations of students' residences. Assign students to closest school. Location/allocation capabilities in Network (ARC/INFO). Digital image provided courtesy of Environmental Systems Research Institute, Inc. Copyright © 1991-1993 Environmental Systems Research Institute, Inc. All rights reserved.

Plate IX. Three dimensional diagram of Scituate Reservoir Watershed; study performed by watershed analysis. Digital image provided courtesy of Environmental Systems Research Institute, Inc. Copyright © 1991-1993 Environmental Systems Research Institute, Inc. All rights reserved.

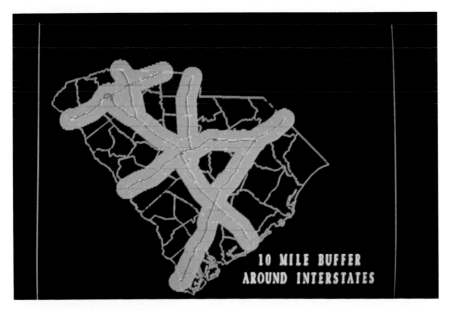

Plate X. Analysis performed on 10 mile buffer from major South Carolina Interstate highways. Digital image provided courtesy of Environmental Systems Research Institute, Inc. Copyright © 1991-1993 Environmental Systems Research Institute, Inc. All rights reserved.

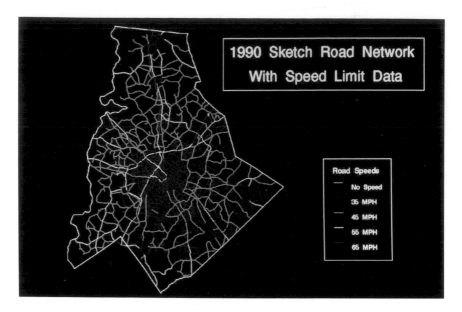

Plate XII. Villages containing endemic Guinea worm infection, Burkina Faso, 1990. Map scale 1:9,395,000. Digital image courtesy of Community Systems Foundation.

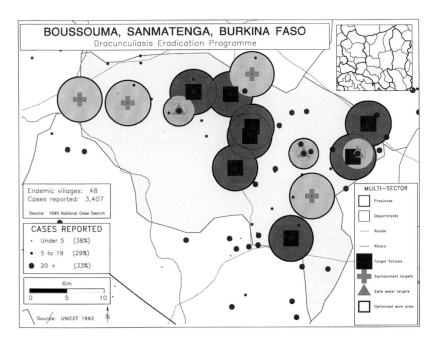

Plate XIII. Villages selected for first round action, UNICEF Guinea Worm Eradication Program, Boussouma Department, Sanmatenga, Burkina Faso. Map scale 1:560,000. Digital image courtesy of Community Systems Foundation.

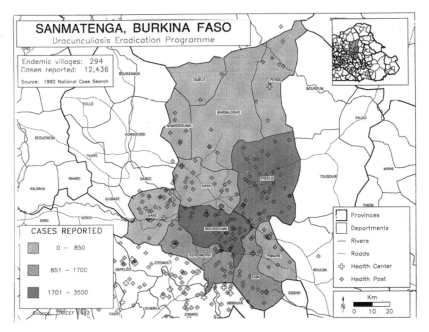

Plate XIV. Dracunculiasis cases reported by Department, Sanmatenga Province, Burkina Faso. Map scale 1:2,916,000. Digital image courtesy of Community Systems Foundation.

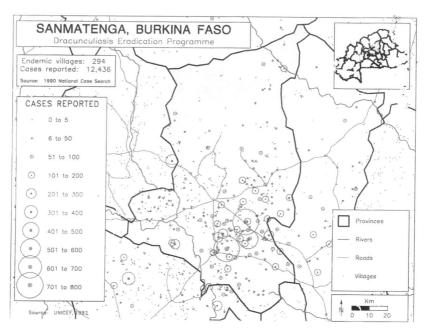

Plate XV. Dracunculiasis cases reported by village, Sanmatenga Province, Burkina Faso. Map scale 1:2,916,000. Digital image courtesy of Community Systems Foundations.

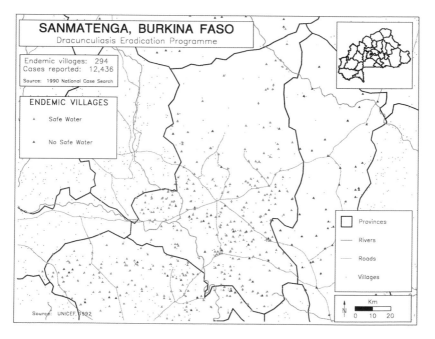

Plate XVI. Water quality in endemic villages, Sanmatenga Province, Burkina Faso. Digital image courtesy of Community Systems Foundation.

PART II: MATERIAL RELATED TO DIGITAL MAPPING

THE UNICEF GUINEA WORM ERADICATION PROGRAM: A GIS APPLICATION

John D. Nystuen,[1]   William D. Drake,[1,2,3]   Halima Dao,[4]

Christine Kolars,[3]   Jean Pierre Meert,[4]   Kris Oswalt,[3]

Michel Saint-Lot,[4]   James M. Sherry[4]

1. Urban and Regional Planning Program,
University of Michigan
2. School of Natural Resources and Environment,
University of Michigan
3. Community Systems Foundation, Ann Arbor, MI. USA
4. United Nations Children's Fund (UNICEF)

When it comes to infectious diseases, health workers know that the environment counts. Certain disease-causing organisms are conveyed from an infected human or animal through an intermediate host, the *vector*, to another human or animal. Vector borne parasites involve the behavior of at least three different organisms, the parasite, the vector, and the host. Knowledge of the behavior and life cycles of organisms coupled with specific data on environmental conditions allows strategies to be formed for disrupting the parasite's life cycle and thus to ridding the human population of the disease. The strength of geographic information systems (GIS) is in the ease with which large volumes of specific details about the environment may be processed. For some infectious diseases specific geographical information is as important in their suppression or eradication as any drug application.

In this chapter an example of an actual public health program, the UNICEF Guinea Worm Eradication Program, in the West African nation of Burkina Faso is presented to demonstrate the use of a geographic information system for program management. The health program is described first to emphasize that using a GIS is a means to an end, a management tool, rather than merely a map display technique although the importance of the displays to program success is not to be minimized. Once the disease and eradication program are described, the chapter details how the GIS was employed to create information from data and what problems arose in carrying out the analysis and fashioning map displays. The example is illustrative but not exhaustive of how a GIS serves as an information management system. The applications are generic and can be used wherever spatial data are involved. The design of the management system, however, is up to the user for only he or she has the necessary knowledge for problem definition.

The UNICEF Guinea Worm Eradication Program

The Guinea Worm disease (Dracunculiasis) is a debilitating but not fatal water-borne parasitic disease endemic to nineteen African countries as well as India and Pakistan. Approximately one million people are infected each year, over 95% of them in Africa. No effective drug treatment for the disease is available. Guinea worm infections are usually limited to one worm at a time, although multiple and repeated infections can occur in the same person. The life cycle of the worm is a key element in its eradication. Because the life cycle is dependent on transmission through a fresh water intermediate host, the tiny, single-eyed cyclops, a crustacean of the subclass Copepoda, disruption at this stage can effectively interrupt the transmission of the worm to humans.

The Guinea worm (Dracunculus medinensis) larvae are released by adult female worms into fresh water where they are ingested by the cyclops. These crustaceans are then ingested by humans when they drink contaminated water. The crustaceans are benign and die once inside the human body, however the guinea worm larvae penetrate the intestinal walls of their human host where they migrate through the viscera, mature and mate. The male worm dies and the female travels through the body to the subcutaneous tissues of the host, usually to the legs and feet, where it grows to full maturity. The adult female is from 60-100 centimeters long. When it is fully mature it burrows through to the outside surface of the human host creating a blister. The blister produces burning and itching sensations, relief from which is often sought by immersing the affected area in cold water. When the blister erupts, the worm emerges through the skin and hundreds of thousands of larvae are discharged into the water where the life cycle begins again as a new generation of larvae are ingested by the cyclops.

Interruption of the life cycle is the basis of the strategy for elimination of dracunculiasis. The life cycle can be interrupted by preventing the infection of drinking water supplies by preventing those with active infections from entering the source, or by surgically extracting the worm (Rohde, et al., 1993); by chemically treating water sources to kill the larvae and/or the intermediate host, cyclops; or preventing the larvae infected cyclops from being ingested by filtering, screening, boiling or chemically treating drinking water.

Four of the major components in the effort to control the guinea worm disease are as follows.

1) **Health education**.

Three messages are conveyed to people living in endemic areas:

(a) the guinea worms come from infected water;

(b) villagers with blisters or emerging worms should not enter sources of drinking water, and should seek safe surgical extraction where it is available;

(c) drinking and food preparation water should be boiled or strained with fine mesh cloth before being consumed (the cyclops is about one millimeter long and easily removed by straining the water).

2) **Provision of potable water**.

Wells and rainwater catchments can be constructed to provide safe water.

3) **Control of the cyclops**.

Use of the larvacide (or cyclopsicide) *abate* is considered effective and safe (due to cost and logistic considerations, vector control is considered complementary to steps one and two).

4) **Monitoring Progress**.

Periodic resurvey of the incidence of the disease.

Used in combination, the four components form an effective disease control strategy and it is due to this feasibility that the guinea worm disease has been targeted for eradication by the WHO/UNICEF Joint Committee on Health Policy (February, 1993) and further endorsed by the Executive Boards of WHO and UNICEF and other bodies.

GIS as a Management Information System

The first step in the eradication program is to acquire information on the prevalence and distribution of the infection along with information about the facilities and organizations (the *infrastructure*) that would be utilized by the program in its operations. The control strategy is largely dependent upon the geographical distribution of this infrastructure. Many endemic villages (villages to which the disease is endemic) are remote and resources to support the control program meager. The goal is to cover as great an endemic area as possible while maximizing the potential of the scarce resources available. Burkina Faso is a low income West African nation without extensive information about population, infrastructure, or environment. The health agencies had to collect much information by direct field survey. In 1990, UNICEF conducted a National Case Search in which 2621 endemic villages and 42,187 cases were reported (UNICEF, 1992). Other data sources were used to locate administrative boundaries, attributes of villages (including their names), location of schools, health clinics, types of water sources, and other infrastructure data. The location of some 13,000 villages throughout the nation had been recorded in machine readable form by a Ministry of Hydrology's nationwide survey of water sources. The locations, recorded as latitude and longitude, were determined by using existing paper maps. A Ministry of Health team while collecting health data for the UNICEF Guinea Worm Eradication Program, employed *global positioning system* (GPS) receivers to pinpoint location of a subset of villages where infections were found. A GPS provides latitude and longitude coordinates through a satellite system link to mobile field instruments, a process that is independent of existing maps. Data from all of these sources constitute the inputs to the GIS management program. In merging these data sets some problems of registration occurred which will be discussed below.

The data were combined to create maps that visually portrayed the prevalence and distribution of infection along with the distribution of schools, health posts, and water points that would be utilized in the eradication program. The process of program design, implementation, and monitoring was divided into three stages called assessment, analysis, and action, a triple-A approach. Here the importance of visual displays cannot be overemphasized. Masses of data specific to thousands of villages in the form of extensive tables and reports are of little use to planners at any level. On the other hand, map displays are able to succinctly present the material at national, provincial, and department (local district) levels in forms easily understood by non-specialists. In the assessment stage posters were constructed which contained maps that portrayed each stage of the program. A national map showing the distribution of the 2621 endemic villages clearly indicated which provinces were most affected (Plate XII). The GIS provides the means to "zoom in" from small national scale to large provincial scale to larger department scale maps in simple operations. This ability greatly assists in the assessment of disease prevalence and in the establishment of priorities for analysis.

In the analysis stage the relationship between disease prevalence and infrastructure was addressed. This was done at district level because it is at this scale that infrastructure and life styles interact. Villages were coded by presence of health centers/posts, schools and by type of water source. Buffers of two and three kilometer radius, assumed to be reasonable walking distances to the facilities, were established around each endemic village that had the appropriate facilities; two kilometers for water sites, three kilometers for health and school locations. The number of cases of dracunculiasis within these buffers were tallied, again an easy operation using a GIS database. Given these details and restricted by limited resources, various combinations of control strategies could be proposed for target districts. Combinations of villages were tried using the overall percent of cases covered as the test of the efficacy of each combination. No formal optimizing method was employed in this procedure but because of the ease of calculation, many combinations were tried and good, if not optimum, strategies discovered. Alternative strategies suggested by different sector managers could also be objectively evaluated one against another. This multisector approach is illustrated in the map of the Boussouma Department (district) in Sanmatenga Province, where, by targeting 4 of 31 health posts and 8 of 14 schools, and adding 3 new safe water points, it was possible to reach 75% of cases reported (Plate XIII).

In other types of problems, the use of formal optimization methods may be appropriate. In such cases the GIS database would be used to create a data set which would be exported to an optimizing program such as a linear programming package with the results then returned to the GIS for visual display.

In the action phase of the policy process, selection of particular infrastructure points for inclusion in the program was based on the level of infection within a department. Villages were classified into highest, middle, and lowest third of villages by number of cases they contain. The top third of endemic villages were targeted for outreach activities including health education administered through schools and construction of safe water points by the Ministry of Hydrology. Villages in the bottom third were targeted for case containment activities administered through local health posts. Case containment consists of monitoring and treatment of infection on a case-by-case basis, a resource intensive undertaking. Case containment is not an indicated action for villages with high levels of infection because case containment in such environments would be swamped by continued re-infection. On the other hand eradication is the goal and where case load is small, a community can be cleared of the parasite by intense case-by-case action. The middle third of the endemic villages did not receive interventions. The expectation was that there would be spill-over benefits from villages having intervention programs through information exchange between villages. Continuous monitoring is planned for the future in order to be able to re-assign priorities after each planning cycle. The entire management plan for the eradication program is substantially strengthened by the GIS database and its timely update.

Critique of Data Processing and Cartographic Displays

The UNICEF Guinea Worm Eradication Program faced many issues common to any GIS database and display effort. The solutions to the data management problems were specific to this program and may or may not be appropriate in other circumstances. This particular application was successful in that decision makers in several national and international agencies as well as local field workers were able to make effective use of the management design and database. As experienced policy makers and managers well know, this is not always the case when trying to use myriad and diverse field reports. The paragraphs below detail data processing and cartographic techniques employed in this application that appear generally useful.

Aggregate Displays

The clarity of a map depends upon the graphic symbolism used. Maps are able to convey quantitative (countable) information as well as general impressions if the symbols are not too large or too numerous as to coalesce. Burkina Faso contains several thousand villages. Each village is represented by a separate record (row) in a spatial database file and each row may contain several attributes (column headings), some village records having more columns than others depending upon availability of information. A large number of choices for displaying all this information is possible. A village is a point object in the GIS database and can be shown on a map as a point symbol. The color and/or shape of the point symbol represents the value of one or another of its attributes. On a national map that is four inches across, as is the case in our examples, the village symbols overlap and are uncountable even when using very small dot symbols (Plate XII). There are over 13,000 green dots and over 2600 red dots displayed in Plate XII. However, the map effectively shows where high concentrations of endemic villages are to be found. Small scale maps displaying large volumes of data are excellent for giving general impressions or overviews of a phenomenon. At the same time, the overview can be backed up by interrogating the database for specific details as needed by changing scale, or creating new subclasses, or performing analysis such as calculating ratios of two variables within spatial units. These operations require use of other spatial units.

The Provinces and Departments are also objects in the spatial database but of a different class. They are two dimensional objects consisting of polygons with closed boundaries. Each village record contains attributes which indicate province and department in which the village is located. These *tags* may be used to relate the different types of spatial objects. For example, the number or proportion of endemic villages by province or department may be calculated and shown in a table or displayed as a *choropleth* map in which extent of infection in departments is indicated by colors or density patterns. In a choropleth map the observational objects are polygons (the provinces or departments) which, in this example, are divided into three classes by number of dracunculiasis cases reported (Plate XIV).

Information at provincial or departmental level is accessible without the need for producing maps at these scales. The GIS database may be used to generate a variety of statistics about each spatial object and the resulting information used in the assessment process. For example, 42,187 cases of Guinea worm infection were detected in 2621 villages, nationwide (20 % of all villages). 12,436 cases (29% of all cases) occur in 294 villages (11% of endemic villages) in the single province of Sanmatenga (1 of 33 provinces (3%)). Which provinces, departments or villages have the highest level of infection are easily determined by rank ordering the spatial units by level of infection and printing a list of their names. The proportions displayed in this fashion make it obvious that Sanmatenga Province warrants selective attention. This procedure was used to target departments and villages in the same manner.

Variable Circle Maps

The distribution of cases reported is also effectively displayed with the use of circles placed on a map at the locations of the endemic villages (Plate XV). The size of a circle represents the class interval to which the village is assigned by number of cases reported. Adjacent symbols overlap; in some instances, the larger ones may completely encircle one or more of the smaller symbols. However, clarity is maintained in crowded areas of the map because the circles are open and overprinting does not obscure the underlying circles. The general effect clearly shows areas of greatest concentration of infection although the number of cases is no longer countable because villages have been grouped into class intervals. That is, the measurement scale has changed from an interval to an ordinal measure. Of course, the underlying GIS database could be used to produce a table which lists the number of village and number of cases reported by class interval. The database could also be used to list by name those villages in each class. Maps are often useful for suggesting further lines of inquiry. Many such inquiries can be pursued by further queries of this sort. Other suggestions might require seeking additional information.

Serendipity: Detecting Spatial Correlates

Graphical displays in general, and maps in particular, are often suggestive of new ideas. To a geographer the map showing the distribution of villages in Burkina Faso has some unusual aspects (Plate XII). Villages are normally clustered along river courses but in south central Burkina Faso there appears to be a zone ten or fifteen kilometers wide on both sides of the rivers that is devoid of villages. This does not seem to hold farther upstream. Here one might suspect another tropical African parasitic disease, *river blindness* (onchocerciasis). The filarial nematode that causes river blindness is a first cousin to the filaria parasites that produce the notorious, disfiguring disease elephantiasis. The microfilariae that causes river blindness causes chronic inflammation of the skin of the victim including eye tissues which can be so damaged as to cause blindness. The vector of the river blindness parasite is a small biting black fly that needs fast flowing, well-oxygenated water containing enough organic material to nourish its filter-feeding larvae. These conditions are normally found along sunlit streams with steeper stream gradients. The fly has a range of ten to fifteen kilometers and the disease is so hazardous as to prevent locating villages in this zone and thus dispossessing people of the potentially fertile agricultural lands within range of infected rivers. This is not a confirmed explanation for the patterns observed on this map but demonstrates the value of maps as a source of ideas for further investigation. If satellite imagery or contour maps were available additional remote analysis would be possible by observing the riverine vegetation and stream gradient of stream courses with and without riverside villages.

The correlate would not have been noticeable if the water courses had not been part of the *backdrop* of the map, if for example, only political boundaries or road networks were used as background elements. The pattern is not discernible when the incidence of guinea worm infection is displayed as a choropleth map (for example Plate XIV). This Plate shows the concentration of infection, but map resolution is less as the spatial units displayed, the two-dimensional departments, are larger than the villages (point objects). Also the water features were not used as backdrop. The association between river courses and village locations is lost. These examples simply emphasize the importance of map design decisions regarding variables to include and symbols to use.

Data Accuracy Revealed by Overlays

The accuracy of a database is always an issue but problems are more likely to surface when using a database that has been compiled from several sources. The Burkina Faso database is a composite which contains some interesting discrepancies.

The river and road data came from the *Carte International du Monde* and *Carte de l'Afrique* digitized at a scale of 1:1,000,000 by the U. S. Geological Survey (USGS) for the USAID Famine Early Warning System (FEWS) in 1985 with an update for the roads in 1990 (Community Systems Foundation). The administrative boundaries were derived from the *IGN Burkina Faso Carte au 1:200,000* and *Carte de l'Afrique de l'Ouest* (1985). Databases containing separate coverages were prepared for several African nations under the FEWS project. To create a map of adjacent nations their common boundaries had to coincide. The separate boundary *strings* (arcs) in the databases of individual nations defining a shared boundary were replaced by the single corresponding boundary string from the comprehensive World Data Bank II (WDBII). This is a 1:3 million scale global database (vector mode) in CD-ROM format from *ArcWorld*, a global database product of the Environmental Systems Research Institute, Inc. (ESRI) of Redlands, California. This database contains national boundaries, along with a great deal of other national level data, and the minor administrative boundaries for some countries but not for Burkina Faso. Hence the FEWS data on interior administrative boundaries were used. The Ministry of Hydrology provided data on the latitude and longitude of the villages. The latitude and longitude coordinates of a subset of the endemic villages came from the Ministry of Health field team's GPS fixes in the field. Inspection of the map reveals discrepancies in spatial data derived from these different sources.

The river meander in south central Sanmatenga shows a lack of correspondence with the provincial boundary (Plates XV, XVI, and Figure 25). The political boundary is offset two to three kilometers from the river and is more simply displayed as a sequence of line segments whereas the river course appears to be a continuous curved line. At at higher resolution the river arcs would be revealed as a string of line segments as well. The political boundaries were digitized using many fewer points per unit length than were used to create the river feature. A trade-off between the spatial resolution and size of the data file is made each time data are digitized from paper maps, photographs, or spatial images. Of course, the resolution is limited in the first instance by the resolution of the source material. The resolution, however, can be generalized by the digitizing operation if curved linear features are traced using a few widely spaced points. The result is a generalized curve stored in fewer lines of file. This procedure may work very well for producing small scale maps such as would be used in a map of the entire nation but it would reveal coarse straight line approximations of curves in large scale maps made from the same data. The decision to digitize the administrative boundaries using widely spaced points may have been dictated by limitations of the GIS software available in 1985.

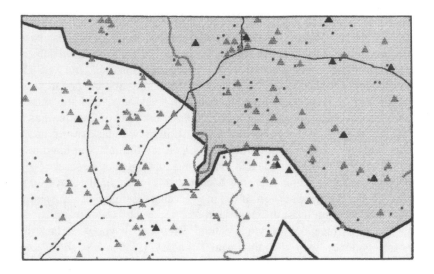

Figure 25. Discrepancy Between Two Databases. Map village water quality, of Sanmatenga Province, Burkina Faso. Legend. Heavy black line -- administrative boundary; thin black line -- road; gray line -- river; endemic villages (gray triangle -- safe water; black triangle -- no safe water); dot -- other villages. Scale: 1:672,000.

When there is a registration problem between two coverages, one might suspect that they were derived from sources using different map projections. This would result in a mismatch and will certainly occur when trying to match locations from air photographs or remote sensing images directly with line maps. Positions of all features in photographs must be corrected to an orthographic plane to account for elevation differences of the terrain in the image, tilt of the aircraft platform, and other known distortions. After adjusting for these systematic distortions the image is projected onto a base using the same map projection as employed on the map to which it is to be registered. Automatic but expensive processes and equipment to accomplish these tasks are available. Conversely, positional data acquired and stored by latitude and longitude graticules refer directly to the curved surface of the globe and do not involve questions of map projections.

Metadata or Marginalia

Paper maps contain entries in the margins of the map, the *metadata* or *marginalia* which are data about the material appearing in the map and the methods of display employed to construct the map. Typically the marginalia includes data acquisition and revision dates, locational grids, map projection, direction of north, scale, map name, map compiler, accuracy levels, and other information useful for establishing proper use of the map. Digital databases should contain files with documentation providing comparable information.

The documentation file accompanying the Generalized Burkina Faso Administrative Boundaries file is an example of metadata and it provides some clues as to the source of the displacement observed on the map. This is the source of the information about the fact that IGN international boundaries were replaced with arcs from WDBII for the administrative boundary coverages. It also indicates that an affine transformation was used to adjust the interior administrative boundaries to fit with the international boundaries that had been replaced with arcs from the WDBII. It further states that the IGN map series uses a UTM projection and that the current coordinates of the file are in decimal degrees. This conversion may be the source of the displacement although this is not clear as unfortunately such complete information is not available regarding the source of the data on the river. All we know is that the river positions are reported in decimal degrees, that is, in longitude and latitude.

Governments, agencies, researchers, and program personnel are creating numerous, often huge and diverse databases on earth phenomena at all scales. Clearly, sharing these data will be essential for effective and efficient data use. Efforts are being made to create the necessary protocol for sharing data (Danko, 1991). The WDBII is a beginning. Local projects should be tied to global databases in systematic fashion. To do this, careful attention must be given to the metadata file and what it need to contain to accomplish that link.

Earth Scale and Field Width

Recording positions by latitude and longitude is recommended as a way of assuring easy integration with other data sets. Problems can be avoided by choosing an appropriate field width for the latitude and longitude records. The meter was originally defined in 1790 as one forty millionth of the circumference of the earth along the meridian passing through Paris. It is still close to that figure even though the standard meter is now measured by an atomic clock and the estimates of the circumference of the earth have changed since 1790. One second of latitude is 1/3600 of a degree or about 31 meters. For use in GIS, latitude and longitude are usually recorded in decimal degrees. One ten thousandth of a degree of latitude is about 11 meters. A good practice is to carry latitude and longitude figures to four or five significant figures to the right of the decimal point if permitted by the accuracy of the data sources. To gain perspective, using a 1:1,000,000 scale paper map source means one millimeter represents one kilometer on the ground. Such a map cannot be used to measure urban lot sizes. Large scale map sources are required to achieve this level of accuracy. GPS technology has this capability. Urban neighborhoods can be mapped by driving around with a GPS receiver and recording location at two second intervals.

In recording latitude and longitude three (or four with sign) significant figures are needed to the left of the decimal point so that field widths of the records need to be seven or eight units wide. If this length is truncated in the data file to yield, for example, one hundredth a degree accuracy (two significant figures to the right of the decimal point), all points will be located on a lattice that is just over a kilometer on the side. This might become noticeable on a large scale map by observing that points appear to line up in approximately kilometer spacing along east-west and north-south axes. The effect will appear if the map has been enlarged well beyond the accuracy of the source.

The scale of Plate XIII is approximately 1:560,000. At this scale, decimal degrees carried to two significant figures to the right of the decimal point would result in all points being located on a lattice of about two millimeter spacing. This would cause a noticeable squaring effect on the map which is not present. Evidently in this case all objects are recorded to at least three significant figures to the right of the decimal point in latitude and longitude graticules.

Error Propagation

In the map shown in Figure 25, both data sets no doubt contain errors and approximations but one is a rather more gross approximation than the other. The displacement is on the order of two to three kilometers. Unfortunately the problem is only apparent at certain locations such as at the river meander feature. The rest of the boundary is at the same level of locational accuracy and this fact may create additional problems.

Villages near the boundary may fall on the wrong side of the line because their positions were derived from other data sources such as the Ministry of Hydrology village file or the GPS fixes from the field teams. Consequently the villages might not be included in analyses of provincial data aggregates which would result in propagation of errors when using provincial totals for calculating ratios or densities or for other purposes. On the other hand, the villages may have been entered into the data set alphabetically by name, province, and department in which case they would be tagged as being in the province (department) regardless of position established by latitude and longitude coordinates. Redundant information regarding spatial set membership can be used to perform error checks. A count of villages tagged to be within a province can be matched against village point locations inside the provincial boundary set (a spatial designation of set membership). Mismatches would be identified. Redundant information is very useful for error checking and should be a design component of a management information system despite the tendency to want to avoid redundancies as a general cost saving principle.

The *zoom in* operation does not change locational accuracy of the underlying database, it simply magnifies the errors along with everything else (Figure 25). The procedure is so easy that greatly enlarged images of data collected at small scale are frequently seen. In this instance, we assume that the political boundary actually coincides with the river course. If we then take the size of the displacement as a measure of positional error or uncertainty, several villages lie within a shorter distance of the apparent location of the boundary than this range of uncertainty. Given this uncertainty, we are not confident we know which province the border villages are in. Furthermore one cannot now be confident of the other maps made from these same data. In Plate XVI, a single department is displayed at a scale of 1:560,000. The western boundary of the department should perhaps be along the river course with which it appears to be intertwined. The same can be said for the road along the northeastern boundary of the department.

Potential errors of this sort cause great distress among local field workers who are likely to have better knowledge of particular local conditions and to realize the data are inaccurate in specific places. This, somewhat unfairly, casts doubt on the validity of the entire data set. The dilemma for researchers is how to deal with local discrepancies. A reasonable strategy is not to use databases compiled at small scale for analysis and then use them for display at too great an enlargement. This is not always an option however, such as in this case, where no other source of departmental political boundaries was available. A reasonable solution for the health planners is to proceed to use the maps, discrepancies and all, but to include an explanation and disclaimer so that less informed users can understand why the mismatches occur. It is not up to the health planners and staff to reconcile national and international databases. Probably the best they can do is to bring problems of this sort to the attention of state authorities who are charged with creating ever-improved national databases.

The boundary problems uncovered did not seriously affect the analysis and management plan of the guinea worm eradication program. For purposes of the project the visible errors on the map were not important. However, other uses of these maps may be quite inappropriate. For example, they should not be used for arguments in boundary disputes.

Use of Default Settings

The use of default settings in GIS operations is the most appropriate way for the causal user to produce acceptable products and displays. Sometimes this works well, sometimes less well. The provincial level and department level maps of Burkina Faso are used as illustrations of this point (Plates XIII and XIV). Notice the insets in the upper right-hand corner of each map. The inset on the provincial map shows the location of the province within the nation. The inset on the department map shows the department's location within the province. The weight of the political boundaries in the inset map is the same as the weight of the political boundaries on the whole map. This is the default condition for making the inset. To achieve a more pleasing effect, the line weights in the inset map should be made to be finer. This could not be done because the particular GIS software used did not have the option of using line weights in the inset that were different from the line weights of the corresponding feature in the main map. The problem does not arise in the departmental map (Plate XIII). The line weights in the inset of the departmental map are quite suitable. The default was used here as well. The default line weight for the provincial boundaries remains the same but as only a few are shown on the inset map and the more abundant departmental boundary lines are rendered as thin lines, the effect does not seem as crowded as is the case in the provincial map.

This seems a small point in the critique of the map displays but the overall artistic impression of a map is made up of many such small points and artistic impression is important for effective communication. Just as with word processors, which make it possible for all sorts of professionals to produce finished manuscripts without the aid of typists, GIS software has made it possible for people other than specialists, i.e., cartographers, to make publishable maps unaided. However, the non-specialist may not be aware of the many small touches a cartographer uses to make a good product. A good policy is to have a map reviewed by a sympathetic critic just as one might have a manuscript reviewed before publishing it. In such a review style as well as content should be considered.

The design options open to the user are a function of the sophistication of the GIS software. Easy-to-use software employs many default settings that the user cannot modify. The more sophisticated systems permit and, unfortunately often require, that the user make many decisions when creating a display. The more sophisticated the system, the harder it is to use. Casual users are better off with simpler, less expensive systems.

Displaying Multiple Attributes

A great deal of information is displayed on the map of the Boussouma Department (Plate XIII). As noted above, each village is a spatial object with several attributes associated with it. Much of the information used in the analysis and action stages of the program are shown on this map including three levels of infection, three types of infrastructure and assumed service areas of villages included in the best program combination. Clarity was maintained by judicious choice of size, shape, color, and overlay sequence of the point symbols used. The size of the symbols relative to the spacing of villages at this scale of map generally avoided overlapping displays of adjacent villages. For instances of multiple occurrences of different attributes for a single village, the symbols are nested with each still visible one inside the other. This results in greater complexity of symbols at points with greater numbers of attributes, a nice touch. The complex, combined symbols must be rather large for easy interpretation. This requirement dictates the map scale used. The result is confusing when villages are too close together as can be seen for the villages in the adjacent department to the south. If these points were part of this particular analysis, an enlargement of this neighborhood would be required.

As a technical matter, the sequence in which the symbols are printed generally needs to be considered because the overprinting of later symbols on earlier ones blanks out earlier symbols wherever they overlap. Thus, first the service area disks are specified, then the larger infrastructure symbols, and finally the level of infection symbols which are the smallest. In this sequence all are visible. Had the sequence been reversed, only the service areas would have been visible.

Need for Backdrop

The villages and their attributes are the *subject* of these maps. Generally the subject of a map display must be presented in the context of a *backdrop* as a means for establishing relative locations of the subject. As here, most often the backdrop consists of political boundaries, road and/or rail nets, and water features. In the Boussouma map the department under study is highlighted to distinguish it from neighboring territories. The details of the backdrop extend outward to the neat line of the map. This is preferable to having the department float on a featureless background. The latter design is often seen, usually dictated by lack of data for the surrounding territories. This is not the case for a database source that includes the entire nation or larger, even worldwide coverage as with the WDBII.

Governments and institutions throughout the world are producing very large, standard databases including hundreds of variables. If these databases contain spatial objects defined as points, lines, or polygons, all the data are subject to spatial analysis and display through GIS programs. GIS software is becoming increasingly sophisticated and user-friendly. Researchers, project developers, program managers, and field workers in both advanced and underdeveloped nations can look forward to increased benefits from the growth of these comprehensive databases and powerful GIS programs.

REFERENCES

1. **Clarke, Keith C., et. al.** The use of remote sensing and geographic information systems in UNICEF's dracunculiasis (Guinea worm) eradication effort. *Preventive Veterinary Medicine*, 11, 229-235, 1991.
2. **Community Systems Foundation.** Mapping Technologies Pack, *GeoTECH*, 1 (1). Prepared for UNICEF by Community Systems Foundation, Ann Arbor, Michigan, March, 1993.
3. **Danko, David M.** The Digital Chart of the World project. *ProceedingsTechnical Papers*, ACSM-ASPRS, Annual Convention, 2, 83-93, 1991.
4. **Desowitz, Robert S.** *New Guinea Tapeworms & Jewish Grandmothers: Tales of Parasites and People.* Avon Books, New York, 1983.
5. **Richards, F., and Hopkins, D. R.** Surveillance: the foundation for control and elimination of Dracunculiasis in Africa. *International Journal of Epidemiology*, 18, 934-943, 1989.
6. **Rohde, J. E., Sharma, B. L., Patton, H., Deegan, C., and Sherry, J. M.** Surgical extraction of Guinea Worm: Disability reduction and contribution to disease control. *American Journal of Tropical Medicine and Hygiene*, 48, 1, 71-76, 1993.
7. **UNICEF** Dracunculiasis Eradication Program Technical Support Team V Meeting Report, Ouagadougou, Burkina Faso, December, 1992.

APPENDIX

DIGITAL MAPPING AND RELATED TECHNOLOGY IN THE FIELD OF INTERNATIONAL HEALTH

Kris Oswalt and Christine Kolars

GIS and other related innovations (such as Global Positioning System receivers) offer a visual means to interrelate situation analysis and program intervention monitoring across sectors. They can also help policy makers to spatially relate the multiple efforts of various collaborating agencies to one another and to specific problems, actions, and outcomes. This means that information pertaining to the health status of an area can be combined with other information, such as number and placement of schools and health posts, in order to create a comprehensive representation of the health situation of a community, district, or nation. Such a comprehensive view can be invaluable as it clearly and easily articulates the parameters of health issues while informing policy makers of the level of infrastructure available to combat health problems. In this vein, GIS and other related technologies have started to be incorporated into the effort to eradicate diseases such as Guinea Worm, and the preliminary assessment of the use of these innovations is highly positive.

SIG et des autres innovations (telles que les récepteurs SPG) offrent la possibilité visuelle de lier l'analyse de situation et la contrôle des interventions à travers les secteurs. Ils peuvent aussi aider les formateurs de la politique à établir le rapport entre les efforts des agences collaboratrices variées et les problèmes, les actions et les résultats spécifiques. C'est à dire que les renseignements appartenant au niveau de santé d'un secteur peuvent être associés avec des autres renseignements, tels que la nombre et lieux d'écoles et de cliniques, pour établir une représentation du niveau de santé de la communauté, de la région, ou du pays. Une telle représentation inappréciable montre les bords des questions de santé et au même temps donne aux formateurs de la politique l'information à propos du niveau d'infrastructure qui existe et qui peut être utilisé à lutter contre des problèmes de santé. Les SIG et des autres technologies associées commencent à être utilisés dans l'effort à éliminer les graves maladies, tel que la maladie du ver de Guinée. L'évaluation initiale de l'utilisation de ces innovations est extrêmement positive.

The use of Geographic Information Systems in health care and disease control is rapidly expanding. Although acceptance of GIS techniques in applied situations is increasing, comprehensive information pertaining to applications of GIS in the field has not been widely available. This scarcity of information has subsequently limited the use of GIS in many sectors. Recently, however, numerous international organizations have voiced an interest in the use of GIS and this, coupled with the desire by health professionals to employ GIS programs well suited to their work, has created the need for technical information tailored to the particular needs and demands of the field of international health.

L'utilisation des SIG s'accroit rapidement. Malgré l'acceptation agrandie du rôle joué par des techniques SIG dans des situations appliquées, les renseignements complets à propos de l'usage de SIG est toujours limités. Ce manque de renseignements limite l'utilisation des SIG dans des secteurs variés. Cedependant, en ce moment beaucoup d'organisations internationales s'intéressent aux applications des SIG et ceci, aussi bien que l'envie des professionels de santé d'employer des logiciels SIG opportuns à leurs metiers, a créé le besoin de renseignements conçus spécialement pour les besoins et les exigences de la domaine de la santé internationale.

Global Positioning System receivers (known as GPS) function on both land and sea and inform users of their precise location, accurate within approximately 25 meters (accuracy varies from product to product). Information such as longitude, latitude, altitude, real-time velocity, and precise time is supplied by the receiver. GPS receivers, relying on signals from orbiting satellites, compare stated and predicted satellite locations in order to calculate distances from the satellite to the user, resulting in an estimation of the user's location.

Les récepteurs du Systèmes Globale de Positionnement (appellés SGP) fonctionnent sur la terre aussi bien que sur la mer et fournissent la position exacte de l'utilisateur, précis à 25 metres (l'exactitude varie de produit à produit). Renseignements tels que la longitude, la latitude, l'altitude, la velocité dans le temps véritable, et l'heure précise sont fournits par le récepteur. Les SGP dépendent sur des signales en provenance des satellites orbitantes, et comparent les positions véritables des satellites avec leurs positions prédirées. Ils calculent les distances des satellites à l'utilisateur et cette distance fournit une estimation de la position de l'utilsateur.

Trials of GPS receivers in the field of international health are beginning to take place. For example, the TransPakII GPS, produced by Trimble Navigation Inc., is currently being used in Burkina Faso as a part of UNICEF's Guinea Worm Eradication Programme.

Des testes appliquées des récepteurs SGP dans la domaine de la santé internationale commencent à être exécutées. Par example, le TransPakII SGP, produit par Trimble Navigation Inc., est utilisé actuellement dans le Programme d'UNICEF d'Eradication du Ver de Guinée dans la Burkina Faso.

ACRONYMS AND ABBREVIATIONS

The digital mapping field is loaded with acronyms; we include a selection, but this is only a start--best advice is to ask for what an acronym stands, if you do not know. Acronyms are only useful when they communicate information; they serve a negative purpose when they intimidate the listener. Acronyms marked with an asterisk are discussed more fully in the encyclopædic part of this handbook.

AAG* - Association of American Geographers
AAT - Arc Attribute Table (ARC/INFO).
ABF - Applications By Forms (INGRES)
ABR* - Automatic Baud Rate detection
AC* - Alternating Current
ACA* - American Cartographic Association
ACAD - AutoCAD (Autodesk, Inc.)
ACD - Automatic Call Distributor
ACES - Austin Communications Education Services
ACHRS - Automated Cable Hazard Reporting System (A.T.&T.)
ACK* - Acknowledgment
ACL - Access Control List
ACM - Association for Computing Machinery
ACS* - Advanced Communications Service
ACSM* - American Congress on Surveying and Mapping
ADCCP* - Advanced Data Communications Control Procedures
ADL - Automated Disk Library (Kodak)
ADM - Add/Drop Multiplexer
ADPCM - Adaptive Quantization Pulse Code Modulation
ADS - Arc Digitizing System (ARC/INFO)
AF - Audio Frequency
AFC - Audio Frequency Control
AGC - Automatic Gain Control
AGI - Association for Geographic Information
AGS* - American Geographical Society
AI* - Artificial Intelligence
AID - Agency for International Development
AIPW - Advanced Image Processing Workstation
AKCLIS - Australian Key Center for Land Information Studies
ALU* - Arithmetic and Logic Unit
ALUF - Analog Line-Up Front
AM - Acquisition Module
AM* - Amplitude Modulation
AM* - Automated Mapping
AM/FM* - Automated Mapping/Facilities Management
AM/FM International - Automated Mapping/Facilities Management International.
AML* - Actual Measured Loss
AML - Arc Macro Language (ARC/INFO)
AMPS* - Advanced Mobile Phone System
AMS - American Mathematical Society

ANDF - Architecture Neutral Distribution Format request for technology
ANI/ALI - Automatic Name Identification/Automatic Location Identification
ANSI* - American National Standards Institute
AOS - Alternate Operator Service
APA - American Planning Association
APD* - Avalanche PhotoDiode
API* - Applications Program Interface
APL* - A Programming Language
APP - Auto Polygon Processing
APPC* - Advanced Program-to-Program Communications
APSRS* - Aerial Photography Summary Record System
APWA - American Public Works Association
ARCNET* - Attached Resources Computing NETwork
ARPANET* - Advanced Research Project Agency NETwork
ARF - Arc cross ReFerence file (ARC/INFO)
ARQ* - Automatic ReQuest for retransmission
ARS - Agricultural Research Service
ARU - Audio Response Unit
ASA - American Standards Association
ASCE - American Society of Civil Engineering
ASCII - American Standards Committee II
ASCII* - American Standard Code for Information Interchange
ASCS - Agricultural Stabilization and Conservation Service
ASD* - Auxiliary Storage Device
ASF - Automatic System Facility
ASIC* - Application Specific Integrated Circuit
ASPRS - American Society for Photogrammetry and Remote Sensing
ASR* - Automatic Send/Receive teleprinter
AT&T - American Telephone & Telegraph
ATDM* - Asynchronous Time-Division Multiplexor
ATM* - Asynchronous Transfer Mode
AURISA - Australian Urban and Regional Information Systems Association
AVA - Automatic Voice Answering
AVC - Automatic Volume Control
AVHRR* - Advanced Very High Resolution Radiometer
AVIRIS* - Advanced Visible/InfraRed Imaging Spectrometer
AVL - Automatic Vehicle Location
AVPO - Axial Vapor Phase Oxidation
AWS - American Wire Gauge
AZP* - Algemeine Zeichen Program
B & S - Brown & Sharpe (gauge)
BASIC* - Beginner's All-purpose Symbolic Instruction Code
BBS - Broadband Switch
BCC* - Block Check Character
B-CDMA* - Broadband Code Division Multiple Access
BCD* - Binary-Coded Decimal
BCNF - Boyce/Codd Normal Form
BELLCORE - BELL COmmunications REsearch
BER* - Bit Error Rate
BERT* - Bit Error Rate Test
BETRS* - Basic Exchange Telecommunications Radio Service
BFL - Back Focal Length
BFO - Beat-Frequency Oscillator

BIA - Bureau of Indian Affairs (USDI)
BIL* - Band Interleaved by Line
BIOS* - Basic Input-Output System
BIP* - Band Interleaved by Pixel
BISDN - Broadband Integrated Services Digital Network
BIT* - BInary ditiT
BITNET* - Because It's Time NETwork
BLERT - BLock Error Rate Test
BLM - Bureau of Land Management (USDI)
BM - Bureau of Mines
BNC* - Bayonet Neill-Concelman
BND - coverage BouNDary file (ARC/INFO)
BOC* - Build-Out Capacitor; Bell Operating Company
BOL* - Build-Out Lattice
BOM - Bill of Materials
BOR - Bureau of Reclamation (USDI)
BPI* - Bits Per Inch
BPNRZ* - BiPolar Non-Return-to-Zero
BPRZ* - BiPolar Return-to-Zero
BPS* - Bits Per Second
BPV* - BiPolar Violation
BRI* - Basic Rate Interface
BSC - Binary Synchronous Communications
BSC* - Binary Synchronous Control*
BSD - Berkeley Software Distribution
BSI - British Standards Institute
BSP* - Bell System Practice
BSQ* - Band Sequential
BTAM* - Basic Telecommunications Access Method
BTL - Backplane Transceiver Logic
BUFR* - Binary Universal Format for data Records
B/W - Black and White
CACM - Communications of the ACM
CAD* - Computer Aided/Assisted Drafting/Drawing/Dispatch
CAD/CAM* - see CAD and CAM
CADD* - Computer Aided/Assisted Drafting and Design
CAE - Computer Aided/Assisted Engineering
CAFM - Computer Aided Facility Management
CAG - Canadian Association of Geographers
CAI - Computer Assisted/Computer Aided instruction
CALS - Computer-aided Acquisition and Logistics Support
CAM* - Common Access Method
CAM* - Computer Assisted Manufacturing
CAM* - Computer Aided Mapping
CAM* - Content Addressable Memory
CAMA* - Computer Assisted Mass Appraisal
CASE - Computer Aided/Assisted Software Engineering
CAT - Cable Analyzer Test
CAT(scan) - Computer Aided Tomography
CATV - Cable Access TeleVision
CATV* - Community Antenna TeleVision
CBD - Commerce Business Daily
CBD - Central Business District

CBS - Compound Boolean Selection
CCA - Canadian Cartographic Association
CCD* - Charge Coupled Device
CCIR - Consultative Committee on International Radio Communications
CCIS - Common Channel Interoffice Signaling
CCITT* - Consultative Committee on International Telephone and Telegraph.
CCRS - Canada Centre for Remote Sensing
CCS* - Common Channel Signaling
CCT* - Computer Compatible Tape
CCTV* - Closed Circuit Television
CCU* - Communications Control Unit
CD* - Carrier Detect
CD - Compact Disk
CDMA* - Code Division Multiple Access
CD-MO - Compact Disk, Magneto-Optical
CD ROM* - Compact Disk Read-Only Memory
CD-WO - Compact Disk, Write Once
CEI - Comparably Efficient Interconnection
CERL - Construction Engineering Research Lab, U.S. Army Corps of Engineers
CEV - Controlled Environment Vault
CF - Composite Feature
CFSRL - Critical Frequency Structural Return Loss
CGA* - Color Graphics Adapter
CGM* - Computer Graphics Metafile
CIA - Central Intelligence Agency
CID* - Charge Injection Device
CIE* - Commission International de l'Eclairage
CIESIN - Consortium for International Earth Science Information Network
CIM - Computer Integrated Manufacturing
CIP* - Change In Plans
CIR* - Color InfraRed (photography or film)
CISC* - Complex Instruction Set Computer
CL* - Cutting Length
CLASS - Custom Local Area Signaling Services
CLLI code - Common Language Location Identifier code
Cm - Centimeter
CMIP* - Common Management Information Protocol
CMIS* - Common Management Information Services
CMOS* - Complementary Metal-Oxide Semi-conductor
CMY* - Cyan-Magenta-Yellow
CO* - Central Office
COBOL - COmmon Business-Oriented Language
CODASYL* - COnference on DAta SYstems Languages
CODEC* - COding and DECoding equipment
COE - U. S. Army Corps Of Engineers; Central Office Equipment
COGO* - COordinate GeOmetry
COM* - Computer Output on Microfilm
COMPANDOR* - COMPressor-expANDOR
COMSEC* - COMmunications SECure
CONECS; CONNECS* - CONnectorized Exchange Cable Splicing
COS* - Corporation of Open Systems
CPE - Customer Premises Equipment
CPI - Common Programming Interface

CPR* - Continuing Property Record
CPS* - Characters Per Second
CPS - Cycles Per Second (now obsolete and replaced by Hertz)
CPU* - Central Processing Unit
CR* - Carriage Return
CRC* - Cyclic Redundancy Checking
CRO - Cathode Ray Oscilloscope
CRRL - Cold Regions Research Lab
CRT* - Cathode Ray Tube
CS - Cursor Stability
CSF - Community Systems Foundation
CSG - Constructive Solid Geometry
CSMA* - Carrier Sense Multiple Access (CSMA/CA, CSMA/CD)
CSOP - Connecting Section Of Plant
CSP - Control Switching Point
CSU* - Channel Service Unit
CT-2* - Cordless Telephone (second generation)
CT-3* - Cordless Telephone (third generation)
CTS* - Clear-To-Send
CTS - Computer Tomography System
CUA - Common User Address
CUG* - Closed User Group
CVD* - Chemical Vapor Deposition process
CVPO - Chemical Vapor Phase Oxidation process
CW - Continuous Wave
CWA - Closed World Assumption
CX - carrier; composite (e.g. signaling-set-leg etc.).
CZCS* - Coastal Zone Color Scanner
D/A* - Digital to Analog Converter
DA - Data Administrator
DAA* - Data Access Arrangement
DAB* - Data Acquisition Board
DACS - Digital Access Crossconnect System
DAP* - Displayable Attribute Processing
DARPA* - Defense Advanced Research Projects Agency
DASD - Direct Access Storage Device
DAT - Digital Audio Tape
dB* - deciBel
DB - DataBase
DBA - Data Base Administrator
DBD - DataBase Description
DBMS* - Data Base Management System
DBRM - DataBase Request Module
DBTG - Data Base Task Group
DC - Direct Current; Data Communications
DCD - Data Carrier Detect. See Carrier Detect
DCE* - Data Communications Equipment
DCL - Digital Command Language
DCS* - Digital Crossconnect System
DCW* - Digital Chart of the World
DD* - DeaD pairs
DDAP - Detailed Distribution Area Plan
DDBMS - Distributed DBMS

DDCMP* - Digital Data Communications Message Protocol
DDGT* - Digital Data Group Terminal
DDL* - Data Definition Language
DDP - Distributed Data Processing
DDR - Data Description Record
DDS* - Dataphone Digital Service
DDS* - Digital Data System
DEA - Drug Enforcement Administration
DEC - Digital Equipment Corporation
DECT* - Digital European Cordless Telephone
DED - Data Element Dictionary
DEDB - Data Entry DataBase
DEF* - Data Exchange Format
DEM - Digital Elevation Matrix
DEM* - Digital Elevation Model
DES - Data Encryption Standard
DEWS - Distribution Engineering Work System (SWBT)
DFAD* - Digital Feature Analysis Data
DFB* - Distributed FeedBack laser
DGIWG - Digital Geographic Information Working Group, NATO
DIGEST - Digital Geographic Information Exchange Standard, DGIWG, NATO
DIME* - Dual Independent Map Encoding
DIMIAS - Digital Interactive Multi-Image Analysis System
DIN* - Deutsche Industrial Norms
DIN - Digital Imaging Network
DIP* - Dual In-line Pins
DLC* - Data Link Control
DLC - Digital Loop Carrier
DLD* - Digital Line Data
DLDS - Digital Line Data Set
DLG* - Digital Line Graph
DLG-E - Digital Line Graph - Enhanced (USGS)
DLMS - Digital Landmass System
DMA [1] - Defense Mapping Agency
DMA [2] - Direct Memory Access
DMI* - Digital Multiplexed Interface
DML* - Data Manipulation Language
DMRS* - Data Management Retrieval System
DN* - Digital Number
DNIC* - Data Network Identification Code
DOC - Department of Commerce
DOD* - Department Of Defense
DOE - Department of Energy
DOS* - Disk Operating System
DOT - Department of Transportation
DP* - Data Processing
DPI - Dots Per Inch
DPSK* - Differential Phase Shift Keying
DRAM - Dynamic Random Access Memory
DRC* - Design Rules Checker
DSA* - Digital Serving Area
DSAS - Digital Subtraction Angiography System
DSC* - Disk Save and Compress

DSDD* - Double-Sided Double Density
DSR - Digital Standard Runoff
DSU* - Data Service Unit
DSX - Digital Signal Crossconnects
DTE* - Data Terminal Equipment
DTED* - Digital Terrain Elevation Data
DTM* - Digital Terrain Model
DTMF* - Dual-Tone Multiple-Frequency
DVI* - DeVice Independent
DXF* - Data Exchange Format
EAS - Extended Area Service
EBCDIC* - Extended Binary Coded Decimal Interchange Code
ECC* - Error Correction Code
ECL - Emitter-Coupled Logic
ECMA - European Computer Manufacturers Association
ECS - Electronic Crossconnect System
EDA - Electronic Design Automation
EDAC - Error Detection And Correction
EDB - Extensional DataBase
EDC - Electronic Data Collection
EDI - Electronic Data Interchange
EDIF - Electronic Design Interchange Format
EDM* - Electronic Distance Measuring
EDP - Electronic Data Processing
EEPLD - Electrically Erasable Programmable Logic Device
EEPROM - Electrically Erasable Programmable Read-Only Memory
EGA* - Enhanced Graphics Adapter
EIA* - Electronic Industries Association
EISA* - Extended Industry Standard Architecture
EL* - Electro-Luminescence
EML* - Expected Measured Loss
EMR* - ElectroMagnetic Radiation
EMS* - Expanded Memory Specification
EMVD - Embedded MVD
ENQ* - ENQuiry
EO cartridge* - Erasable Optical cartridge
EO drive* - Erasable Optical drive
EOC - Embedded Operations Channel
EOF* - End Of File
EOL - End Of Line
EOT* - End Of Text
EOSAT* - Earth Observation SATellite company
EPA - Environmental Protection Agency
EPLD - Erasable Programmable Logic Device
EPROM* - Erasable Programmable Read Only Memory
EQUEL - Embedded QUEL
ERDAS - Earth Resources Data Analysis System (ERDAS)
ERIM - Environmental Research Institute of Michigan
ERLL - Enhanced Run Length Limited
EROM - Erasable Read Only Memory
ERTS-1* - Earth Resources Technology Satellite-1
ESA - European Space Agency
ESD - Engineering Society of Detroit

ESDI* - Enhanced Small Device Interface
ESIO - Earth Science Information Office
ESP - Enhanced Service Provider
ESRI, Inc. - Environmental Systems Research Institute, Inc.
ETL - Engineer Topographic Lab (U.S. Army)
ETX* - End of TeXt
EWO - Engineering Work Order
4GL* - Fourth-Generation Language
FAA - Federal Aviation Administration
FCC* - Federal Communications Commission
FCIF - Flexible Computer Interface Format
FCS* - Frame Check Sequence
FD - Functional Dependence
FDDI* - Fiber Distributed Data Interface
FDHD* - Floppy Drive High Density
FDM* - Frequency Division Multiplex
FDMA* - Frequency Division Multiplexing Architecture/Multiple Access
FEMA - Federal Emergency Management Agency
FEP - Front End Processor
FERC - Federal Energy Regulatory Commission
FET* - Field-Effect Transistor
FEWS - Famine Early Warning System (USAID)
FF* - Form Feed
FGCC* - Federal Geodetic Control Committee
FHA - Federal Housing Administration
FHWA - Federal Highway Administration (DOT)
FIC-CDC - Federal Interagency Coordinating Committee on Digital Cartography
FIGS* - *FIGures Shift*
FIP - File Inquiry Processor
FIPS* - Federal Information Processing Standard
FL - Focal Length
FM* - Frequency Modulation
FMS/AC - Facility Mapping System/AutoCAD
FORTRAN* - FORmula TRANslation
FOV* - Field Of View
FPGA* - Field-Programmable Gate Array
FRAMME - Facilities Rulebased Application Model Management Environment (Intergraph)
FRC* - Field Reporting Code
FS - U. S. Forest Service
FSK* - Frequency Shift Keying
FTAM* - File Transfer, Access, and Management
FTP* - File Transfer Protocol
FWS - U. S. Fish and Wildlife Service (USDI)
FY - Fiscal Year
GaAS* - Gallium Arsenide
GAIM - Global Analysis, Interpretation and Modelling (IGBP)
GB* - GigaByte--10^9 bytes
GBF/DIME* - Geographic Base File/Dual Independent Map Encoding
GBIS - Geo-Based Information System
Gbps* - Gigabits per second
GCDB - Geographic Coordinate Data Base

GDI - Geographic Data Integrator
GEMS - Global Environmental Monitoring System (UNEP)
Geo/SQL - Geographic/Structured Query Language
*GFI** - Group Format Identifier
GFIS - Geographic Facilities Information System
*GHz** - GigaHertz
GIF - Graphic Interchange Format
*GIGO** - Garbage In, Garbage Out
GIMMS - Geographic Information Manipulation and Mapping Systems
*GIRAS** - Geographic Information Retrieval and Analysis System
*GIS** - Geographic Information System
*GHz** - GigaHertz
*GKS** - Graphics Kernel System
GNIS - Geographic Names Information System
*GOS** - Grade Of Service
GOSIP - Government Open Systems Interconnect Profile
GPIB - General Purpose Interface Board
*GPS** - Global (or Geodetic) Positioning System
GRASS - Geographic Resources Analysis Support System
GRID - Global Resource Information Database (UN Environment Program)
GSA - General Services Administration
GSFC - Goddard Space Flight Center
*GUI** - Graphical User Interface
*HD** - High Density diskette
HD/GEC - Human Dimensions of Global Environmental Change Programme (ISSC)
*HDDT** - High-Density Digital Tape
*HDLC** - High-level Data Link Control
HDTV - High Definition TeleVision
*HDX** - Half-DupleX transmission
*HFSP** - Human Frontiers Science Program
HHS - U. S. Department of Health and Human Services.
HIS - Hospital Information System
*HIS** - Hue, Intensity, and Saturation
*HLS** - Hue, Luminance, and Saturation
*HMA** - High Memory Area
HOM - Hotline Oblique Mercator
*HPGL** - Hewlett Packard Graphics Language
HUD - U. S. Department of Housing and Urban Development
HVAC - Heating Ventilation Air Conditioning
*Hz** - Hertz
*I/O** - Input/Output
IBG - Institute of British Geographers
IBIS - Image-Based Information Systems
IBM - International Business Machines Corporation
*IC** - Integrated Circuit
ICA - International Cartographic Association
ICAM - Integrated Computer Automated Manufacturing
ICC - Interstate Commerce Commission
ICSU - International Council of Scientific Unions
ICS/UM - Interactive Communications and Simulations, University of Michigan (School
　　　of Education)
ID - Identification
*ID** - Idle Dead

IDB - Intensional DataBase
IDE - Integrated Drive Electronics
IDLC - Integrated Digital Loop Carrier
IDMS - Integrated DataBase Management System
IDS - Image Display System
IEEE* - Institute of Electrical and Electronic Engineers
IFIAS - International Federation of Institutes of Advanced Studies
IFOV* - Instantaneous Field of View
IGBP - International Geosphere-Biosphere Programme
IGBP-DIS - *IGBP* Data and Information System
IGDS - Interactive Graphic Design System (Intergraph)
IGES* - International Graphic Exchange Specification
IGIS - Integrated Geographic Information System (Intergraph)
IIP - Intercept Inquiry Processor
ILD* - Injection Laser Diode
IMaGe - Institute of Mathematical Geography
IMP - Intercept Maintenance Processor
IMS - Information Management System
IMTS* - Improved Mobile Telephone Service
INGRES - INteractive Graphics and REtrieval System
INMD - In-Service, Non-Intrusive Measurement Device
IOS - International Organization Standardization
IPI* - Intelligent Peripheral Device
IPID* - Item of Plant IDentifier
IPS - Image Processing System
IPX* - Internetwork Protocol eXchange
IRG* - Inter-Record Gap
IRM* - Information Resource Management
IRQ* - Interrupt ReQuest lines
IS - Information Systems
ISA - Industry Standard Architecture
ISAM - Index Sequential Access Method
ISDN* - Integrated Services Digital Network
ISO* - International Standards Organization
ISPLS - Indiana Society of Professional Land Surveyors
ISS - Image Storage System
ISSC - International Social Science Council
ITC - International Training Center for Aerospace Survey and Earth Sciences (The Netherlands)
ITD - Institute for Technology Development
ITU* - International Telecommunications Union
ITUM - Integrated Terrain Unit Mapping
IUFRO - International Union of Forestry Research Organizations
IXC - Inter Exchange Carrier
JD - Join Dependence
JGOFS - Joint Global Ocean Flux Study (SCOR/IGBP)
JMOS - Job Management Operations Systems (Bell)
JNC* - Jet Navigation Charts
JPL - Jet Propulsion Laboratory
JSC - Johnson Space Center
KB* - KiloByte--10^3 bytes
Kbps* - Kilobits per second

KHz* - KiloHertz
KSR* - Keyboard Send/Receive
LAD - Local Area Data channel
LAN* - Local Area Network
LANDSAT - U. S. Satellite System
LAP* - Line Access Procedure
LAPB* - Line Access Procedure, Balanced
LAPD* - Link Access Procedure for the D-channel
LAPM* - Link Access Procedure for Modems
LASER* - Light Amplification by Stimulated Emission of Radiation
LATA* - Local Access and Transport Area
LAWN - Local Area Wireless Network
LCD* - Liquid Crystal Display
LCGU - Least Common Geographic Unit
LDF - Light Distribution Frame; Lightwave
LEC - Local Exchange Carrier (or Company)
LED* - Light Emitting Diode
LEF* - Local Event Flag
LEIM - Loop Electronic Inventory Module (Bell)
LEIS - Loop Engineering Information Systems (Bell)
LEN - Local Exchange Node
LF* - Line Feed
LFACS - Loop Facility Assignment and Control System (Bell)
LIDAR - LIght Detection And Ranging
LIS* - Land Information System
LNA - Low Noise Amplifier
LNB - Low Noise Block
LNC - Low Noise Converter
LOC - Library of Congress
LOICZ - Land-Ocean Interactions in the Coastal Zone (IGBP, Proposed)
LON - Local Operating Network
LP - Line Processor
LRC* - Longitudinal Redundancy Check
LRS* - Land Records System
LSI* - Large Scale Integration
LTRS* - LeTteRs Shift
LU 6.2* - Logical Unit 6.2
MAC* - Media Access Control
MAN - Metropolitan Area Network
MAP* - Manufacturing Automation Protocol
MAP* - Map Analysis Package
MB* - MegaByte
Mbps* - Megabits per second
MC&G - Mapping, Charting, and Geodesy
MC&GFDES - MC&G Feature Data Exchange Structure (DMA)
MCF - Million Conductor Feet
MCP - Multiplex and Cross-connect Processor
MCVD* - Modified Chemical Vapor deposition Process
MDF* - Main Distribution Frame
MDT - Mobile Data Terminal
MFJ* - Modified Final Judgment
MFLOPS* - Million FLOating Point operations per Second
MFM* - Modified Frequency Modulation

MHz* - MegaHertz
MIC - Material Identification Code
MIPS - Microimage Processing System
MIPS* - Million Instructions Per Second
MIS - Management Information System
MMS - Minerals Management Service
MMU - Memory Management Unit
MNP - Microcom Networking Protocol
MODEM* - MOdulator-DEModulator
MOESI* - Modified, Owned, Exclusive, Shared, Invalid
MOS* - Metal Oxide Semiconductor
MOSS - Map Overlay and Statistical System
MOTIS - Message-Oriented Text Interchange System
MOU - Memorandum of Understanding
MPL - Multi-Schedule Private Line
MRIS - Magnetic Resonance Imaging System
MRP - Manufacturing Resource Planning
MSA* - Metropolitan Statistical Area
MSI* - Medium Scale Integration
MSS - MultiSpectral Scanner
MTBF* - Mean Time Between Failures
MTTR* - Mean Time To Repair
MULDEX* - MULtiplex/DemultiplEX
MUX* - MUltipleXor
MVD - MultiValued Dependence
MVTEX (wait) - MUTual EXclusion state (DEC)
NA* - Numerical Aperture
NAK* - Negative AcKnowledgment
NAPLPS* - North American Presentation Level Protocol Syntax
NAPP aerial photographs* - National Aerial Photography Program aerial photographs
NARS - National Archives and Records Service
NASA* - National Aeronautics and Space Administration
NASS - National Agricultural Statistics Service
NATO - North Atlantic Treaty Organization
NBS* - National Bureau of Standards
NCDCDS - National Committee for Digital Cartographic Data Standards
NCGA - National Computer Graphics Association
NCGIA* - National Center for Geographic Information and Analysis
NCIC - National Cartographic Information Center (USGS)
ND* - Normalized Difference
NDCDB - National Digital Cartographic Data Base
NETBIOS - NETwork Basic Input/Output System
NFS* - Network File Server
NGRS - National Geodetic Reference System
NGS - National Geographic Society
NHAP* - National High Altitude Program
NIC* - National Interface Card
NIST* - National Institute of Standards and Technology (formerly NBS)
nm* - nanometer
NMS - Nuclear Medicine System
NNF - Nested Normal Form
NOAA - National Oceanic and Atmospheric Administration
NOAA/NESDIS - National Oceanographic and Atmospheric Administration/National

Environmental Satellite, Data and Information Service

NOAA/NGDC - National Oceanographic and Atmospheric Administration/National Geophysical Data Center

NOAA/NMFS - National Oceanographic and Atmospheric Administration/National Marine Fisheries Service

NORDA - Naval Ocean Research and Development Activity (U. S. Navy)

NOS - National Oceanic Survey

NOS* - Network Operating System

NOSC - Naval Ocean Systems Center

NPS - National Park Service (of the USDI)

NRC - Nuclear Regulatory Commission; National Research Council

NRZ* - Non-Return to Zero

NRZI* - Non-Return to Zero, Inverted

NSF - National Science Foundation

NSTL - National Space Technology Lab (now SSC)

NTIA* - National Telecommunications and Information Agency

NTN - Network Terminal Number

NTSC - National Television Standards Committee

NUI - Network User Identification

NURBS - Non-Uniform Rational B-Splines

OC - Optical Carrier

OCC - Other Common Carrier

OCR scanner - Optical Character Recognition scanner

OCWR - Optical Continuous Wave Reflectometry (Reflectometer)

OEM - Original Equipment Manufacturer

OFDR - Optical Frequency Domain Reflectometry (Reflectometer)

OLTP - On Line Transaction Processing

OMB - U. S. Office of Management and Budget

ONA - Open Network Architecture

ONC* - Operational Navigation Charts

ONI - Optical Network Interface

ONR - Office of Naval Research

OOPS* - Object-Oriented Programming System

OPIS- Outside Plant Information System (Synercom)

ORNL - Oak Ridge National Laboratory

OS - Operating System

OS/2 - Operating System/2

OSD - On Site Demonstration

OSF* - Open Software Foundation

OSI* - Open Systems Interconnection

OSL - Operation Specific Language

OSMRE - Office of Surface Mining Reclamation and Enforcement

OSP - Optical Storage Processor

OSP* - OutSide Plant

OSS - Operational Support Systems

OTDR - Optical Time Domain Reflectometry (Reflectometer)

P-N junction* - Positive-Negative junction

PABX* - Private Automatic Branch eXchange

PAC* - Plant Account Code

PACS - Picture Archiving and Communication System

PAD* - Packet Assembler/Disassembler

PAL* - Programmable Array Logic

PAM* - Pulse Amplitude Modulation

PAMA* - Pulse Address Multiple Access
PBX* - Private Branch Exchange
PC* - Personal Computer
PCB - Printed Circuit Board; Program Communication Block
PCM* - Plug Compatible Manufacturer
PCM* - Pulse Code Modulation
PCMCIA - Personal Computer Memory Card International Association
PCN* - Personal Communications Network
PCP* - Preferred Count Primary pairs
PCS* - Permanent Secondary Pairs
PCS* - Personal Communications Service
PCS* - Preferred Count Secondary pairs
PD* - PhotoDetector
PDES* - Produce Data Exchange Specification
PDL - Page-Description Language
PDN* - Packet Data Network
PDN* - Public Data Network
PGS* - Pair-Gain Systems
PHIGS* - Programmer's Hierarchical Interactive Graphics System
PI - Photo Interpretation
PIC - Primary Interexchange Carrier
PIC* - Plastic Insulated Conductor
PIN* - Parcel Identification Number
PIN diode* - Positive-Intrinsic-Negative diode
PIP - Peripheral Interchange Program
PLA - Programmable Logic Array
PLD - Programmable Logic Device
PLR* - Plant Location Record
PLRMS - Plant Location Record Management System (Southern Bell Telephone)
PLSS* - United States Public Land Survey System
PM* - Prime Meridian
POST* - Power-On System Test
POTS* - Plain Old Telephone Service
PP - Path Processor
PP* - Pre-Posting
PPI - Plan Position Indicator
PPP* - Point-to-Point Protocol
PRESS - Plant Records Engineering Scanning System (Bell of Pennsylvania)
PRF* - Plat Request Form
PRI* - Primary Rate Interface
PROM* - Programmable Read Only Memory
PSDN* - Packet-Switched Data Network
PSK* - Phase Shift Keying
PSTN - Public Switched Telephone Network
PTT* - Post, Telephone, and Telegraph authority
PVC* - Permanent Virtual Circuit
PVC - PolyVinyl Chloride
QA - Quality Assurance
QAM* - Quadrature Amplitude Modulation
QAM - Quality Assessment Measurement.
QBE - Query By Example.
QBF - Query By Forms
QC - Quality Control

QIC* - Quarter-Inch Compatibility
QMF - Query Management Facility
QUEL - QUEry Language
RADAR - RAdio Detection And Ranging
RAID - Redundant Array of Independent Disks
RAM* - Random Access Memory
RBF - Report By Forms
RBHC - Regional Bell Holding Company
RBOC - Regional Bell Operating Company
RBV* - Return Beam Vidicon
RD - Received Data
RDB* - Relational Database
R+DBMS - Relational Data Base Management System
RDMS - Relational Database Management System
RF - Radio Frequency.
RFI* - Radio Frequency Interference
RFI* - Request For Information
RFP* - Request For Proposal
RFQ* - Request For Quote/Quotation
RGB* - Red-Green-Blue
RHC - Regional [Bell] Holding Company.
RICS* - Re-enterable In Cable Splice
RIFF - Raster Image File Format.
RIP - Retired In Place.
RIS - Radiology Information System
RISC* - Reduced Instruction Set Computing
RJE* - Remote Job Entry
RLL* - Run-Length Limited
RLSD - Received Line Signal Detector
RMS - Root Mean Square
RMSE - Root Mean Square Error.
RN - Remote Node
RO* - Read-Only
ROM* - Read-Only Memory
RPG* - Report Program Generator language
RPM - Revolutions Per Minute.
RR - Repeatable Read
RRC - Regional Research Centre
RRN - Regional Research Network
RRS - Regional Research Site
RS - Remote Sensing
RSA - Regional Science Association
RT* - Remote Terminal
RTL - Run-Time Library
RTS* - Request To Send
SAA* - System Application Architecture (IBM)
SAC - Scientific Advisory Council (IGBP); Strategic Air Command
SAC* - Serving Area Concept
SAG* - Street Address Guide
SAI* - Serving Area Interface
SAN - Small Area Network
SAR* - Synthetic Aperture Radar
SCA - Suitability/Capability Analysis

SC-IGBP - Special Committee for the IGBP (supplanted in September 1990 by Scientific Committee for the IGBP)

SCO - Santa Cruz Organization/Operation

SCOR - Scientific Committee on Ocean Research

SCPC - Single-user Channel Per Carrier.

SCS - Soil Conservation Service

SCSI* - Small Computer Systems Interface

SDD-1 - System for Distributed Databases-1

SDH - Synchronous Digital Hierarchy

SDIAO - Stereo Digitizing Interface Absolute Orientation

SDLC* - Synchronous Data Link Control

SDM - Space Division Multiplex.

SDTS - Spatial Data Transfer Standard, Federal Geographic Data Committee

SEED* - Self-Electro-optic-Effect-Device

SEQUEL* - Structured English QUEry Language

SIF* - Standard Interchange Format

SIG - Street Index Guide.

SIR - Shuttle Imaging Radar.

SLC* - Subscriber Loop Carrier

SLD - SuperLuminescent Diode

SLF - Standard Linear Format

SLIP* - Serial Line Internet Protocol

SMA* - Standard Metropolitan Area

SMB* - Sewer Message Block

SMD* - Storage Module Device

SME - Subject Matter Expert (Bell)

SMF* - Single Mode Fiber

SMF/DS* - Single Mode Fiber/Dispersion Shifted

SMP - Symmetrical MultiProcessing (DEC)

SMR - Specialized Mobile Radio

SMRT* - Single Message-unit Rate Timing

SMSA* - Standard Metropolitan Statistical Area

S/N* - Signal-to-Noise ratio

SNA* - Systems Network Architecture

SNMP* - Simple Network Management Protocol

SNR* - Signal-to-Noise Ratio

SOH - Start Of Header

SONET* - Synchronous Optical NETwork

SP - Section Processor

SPARC* - Scalable Processor ARChitecture

SPC* - State Plane Coordinates

SPCS* - State Plane Coordinate System

SPE - Synchronous Payload Envelope

SPOT* - Système Probatoire d'Observation de la Terre.

SPSE - Society of Photographic Scientists and Engineers

SQL* - Structured Query Language

SRAM - Static RAM

SRDM* - Sub-Rate Data Multiplexer

SSA - Segment Search Argument

SSB* - Single Side Band

SSC - Stennis Space Center (formerly NSTL)

SSD - Solid-State Disks

SSDD diskette - Single Sided Double Density diskette

SSI* - Small-Scale Integration
ST - Service Termination
START - Global Change System for Analysis, Research and Training
STDM - Statistical Time Division Multiplexing
STM - Synchronous Transfer Mode
STS - Shared Tenant Services
STX - Start of TeXt.
SVC - Switched Virtual Circuit
SVGA* - Super Video Graphics Array
SWCC - Second World Climate Conference (1990)
SYMAP* - SYnagraphic MAPping program
SYSGEN - SYStem GENeration utility
TARGA file* - Truevision Advanced Raster Graphics Adapter file

TB* - TeraByte--10^{12} bytes
TCAC* - T-Carrier Administration Center
TCAM - Telecommunication Access Method
TCP/IP* - Transmission Control Protocol/Internet Protocol
TCU - Transmission Control Unit.
TD - Transmitted Data.
TDM* - Time Division Multiplexing
TDMA* - Time Division Multiplexing Architecture; Time Division Multiple Access.
TELCO - TELephone COmpany
TELEX* - TELeprinter EXchange service
TFIDS - Telephone Facilities Interactive Data System (Intergraph)
TFT - Thin Film Transistor
TIA* - Telephone Industry Association
TIF* - Telephone Influence Factor
TIFF* - Tagged Image File Format
TIGER (files)* - Topologically Integrated Geographic Encoding and Reference files
TIN* - Triangulated Irregular Network
TM - Registered Trademark
TM* - Thematic Mapper (on LANDSAT satellite)
TNC connector - Threaded Neill-Concelman connector
TOP* - Technical and Office Protocol
TPC - Telecommunications Process Controller
TPI* - Tracks Per Inch
TSAT - T-carrier Small Aperture Terminal
TSO - Time-Sharing Option
TSR - Terminate and Stay Resident
TTY - TeleTYpewriter
TUG - TeX Users' Group
TVA - Tennessee Valley Authority
TVI* - Transformed Vegetation Index
TWX* - TeletypeWriter eXchange service
UAF - User Authorization File
UART* - Universal Asynchronous Receiver Transmitter
UCI - Unified Customer Interface
UDL - Unified Data Language
UGISA - University Geographic Information System Alliance
UHF* - Ultra High Frequency
UIC* - User Identification Code
ULSI* - Ultra Large Scale Integration

UN - United Nations
UNCED - United Nations Conference on Environment and Development
UNEP - United Nations Environment Program
UNESCO - United Nations Educational, Scientific and Cultural Organization
UNI - User-Network Interface
UNICEF - United Nations International Children's Emergency Fund
UNIX - A multi-user operating system
UNU - United Nations University
UPS* - Uninterruptable Power Supply
URISA - Urban and Regional Information System Association
URP/GIS/UM - Urban & Regional Planning GIS lab, Univ. Michigan
US - United States; Ultrasound.
USAF - U. S. Air Force
USAID - U. S. Agency for International Development
USASCII* - U.S. ASCII
USDA - U. S. Department of Agriculture
USDI - U. S. Department of the Interior
USFS - U. S. Forest Service
USFWS - U.S. Fish and Wildlife Service (of the USDI)
USGS - U. S. Geological Survey (of the USDI)
USPLS - U. S. Public Land Survey
USS - Ultra Sound System
USTA - United States Telephone Association
USTSA - United States Telephone Suppliers Association
UTM (grid)* - Universal Transverse Mercator grid
UTM (projection)* - Universal Transverse Mercator projection
UTP* - Unshielded Twisted Pair
UV* - Ultraviolet
UWA - User Work Area
VANS - Value Added Network Services.
VAR - Value Added Reseller
VAX - Virtual Address eXtension (DEC)
VDT* - Video Display Terminal
VDU* - Visual Display Unit
VESA - Video Electronics Standards Association
VF - Voice Frequency
VGA* - Video Graphics Array
VHDL - VHSIC Hardware Description Language
VHF* - Very High Frequency.
VHS - Video Home System
VHSIC - Very High Speed Integrated Circuit
VICAR - Video Information Communication and Retrieval
VIFRED - VIsual FoRms EDitor
VIGRAPH - VIsual GRAPHics
VLSI* - Very Large-Scale Integration
VNL* - Via Net Loss
VNLF* - Via Net Loss Factor
VOxEL* - VOlume ELement
VPF* - Vector Product Format
VPN - Virtual Private Network
VRAM - Video (dual-ported) Random Access Memory.
VSAT* - Very Small Aperture Terminal
VT - Virtual Tributary

WAN* - Wide Area Network
WATS - Wide Area Telephone Service
WCP - World Climate Programme
WCRP - World Climate Research Programme (*WMO/ICSU*)
WCRP-JSC - Joint Scientific Committee (*WCRP*)
WDM* - Wavelength Division Multiplexing
WFF - Well-Formed Formula
WGS* - World Geodetic System
WHO - World Health Organization
WIMP* - Windows, Icons, Mouse, Pull-down menus
WMO - World Meteorological Organization
WOCE - World Ocean Circulation Experiment (*WCRP*)
WOL* - Wired Out of Limits
WOM - Work Order Manager
WORM* - Write Once, Read Many
WPG* - Work Print Generation
WRS - Worldwide Reference System
WYSIWYG* - What You See Is What You Get
XD - indeXeD
XGA* - Extended Graphics Adapter
XNS* - Xerox Network Services
XNS/ITP - Xerox Network Systems' Internet Transport Protocol
X3H2 - ANSI database committee
YMCK* - Yellow-Magenta-Cyan-Black
ZTS* - Zoom Transfer Scope

INTELLIGENT VEHICLE-HIGHWAY SYSTEMS
TERMS AND CONCEPTS

Up-to-date reports of current traffic conditions may help motorists to avoid rush-hour accidents in major cities by suggesting alternate routes around existing congested areas; signs at the edge of the road may inform travelers of hazards as well as conveniences. There are numerous other ways that travelers get information as they drive, all of which are related to the level of technology to which they have access.

Research efforts in the United States, Western Europe, and Japan over the past thirty years at using new technology to create "smart" interactive pairings of road, driver, and vehicle offer exciting glimpses of surface travel in the future. These "systems" which use electronic and other state-of-the-art technology are "intelligent vehicle-highway systems" (IVHS).

Thus, the future might bring countinuous routing instruction with a digital map, that interacts with data concerning current traffic conditions, displayed within a vehicle in the driver's line-of-sight. Digital maps combined with dead reckoning devices attached to wheel pickups might off automatic vehicle navigation. Motorist information systems, using radio signals might be coupled with digital mapping technology so that both audio and visual information comes into the car from buried cable antenna in the road, from roadside television camers and data links at the road side, or from various navigation beams including from remotely sensed satellite systems.

Automatic vehicle location, automatic vehicle identification and billing, automatic tolls and road pricing, obstacle detection systems, radar sensing of road markings, collision warnings and avoidance systems, as well as automated highways that platoon vehicles with regard to spacing, and vehicle and road-edge detection--almost as if the vehicles were grocery items spread out on a supermarket check-out lane conveyor belt--are only a few of the exciting possibilities just around the corner. Progress on this topic has involved considerable research in academic institutions, private industry, and government; new words, phrases, and acronyms surround the communication of IVHS ideas; funding initiatives and political involvement are prevalent from the local to the federal levels. In the near future, there will be numerous market opportunities for individuals with background in IVHS-related topics; some estimates suggest market opportunities of over \$210 billion, in research and development, field tests, and deployment of IVHS, over the next twenty years. The organization of information has become a significant task.

One way to organize various components of IVHS is according to "functional areas" suggested by Chen and Ervin in The University of Michigan's IVHS Seminar. They are:

ATMS: Advanced Traffic Management Systems
  including electronic toll collection, traffic surveillance, and adaptive signal light control;
ATIS: Advanced Traveler Information Systems
  including motorist information, pre-trip planning, and dynamic route guidance;
AVCS: Advanced Vehicle Control Systems
  including intelligent cruise control, lane sensing and lateral control, and platooning;
CVO: Commercial Vehicle Operation
  including weigh-in-motion for trucks, automatic vehicle classification, and border crossing inspection;

APTS: Advanced Public Transportation Systems
  including automatic vehicle location, smart cards as super transfers, and dynamic ride sharing;
ARTS: Advanced Rural Transportation Systems
  including collision avoidance, enhanced vision, and autonomous navigation system.

What follows is an alphabetical list of terms and acronyms, and concepts concerning IVHS. This listing is only a beginning; it is all changing very quickly --- at best, we hope that this will be helpful as an entry point to the terminology employed in the late 1980s and early 1990s.

AACS - Automobile-Automobile Communication Systems (Japan).
ADIS - see ATIS.
ADVANCE - Advanced Driver & Vehicle Advisory Navigation Concept; system in the Chicago area suburb arterials covering 250 square miles using a variety of current technologies. (Motorola.)
AEDIC - Advanced Easy Driving and Informatic Car.
AHS - Automated Highway Systems.
ALI - Autofarer Leitung und Informationsystem Program; European; 1979.
ALI-SCOUT--Berlin IVHS;. (Siemens.)
AMTICS - Advanced Mobile Traffic Information and Communication System; Japanese system; 1987.
APTS - Advanced Public Transportation Systems.
ARTS - Advanced Rural Transportation Systems.
AST - Active Safety Technologies.
ATCS - Area Traffic Control and Surveillance System.
ATIS - Advanced Travelers Information Systems (replaces ADIS - Advanced Drivers Information Systems).
ATMS - Advanced Traffic Management Systems.
AVC - Automatic Vehicle Classification.
AVCS - Advanced Vehicle Control Systems.
AVI - Automatic Vehicle Identification.
BEC - Benefits, Evaluation, and Costs.
CACS - Comprehensive Automobile Traffic Control System, Japan, 1976.
Caltrans - California Department of Transportation; this department has been a leader in promoting this sort of technology.
Cellular phones - used as location probes in cars in some systems.
Crescent - see HELP.
CSMA - Carrier Sense Multiple Access.
CVO - Commercial Vehicle Operations.
Dead-reckoning - the process of keeping a running account of changes in position from which to estimate new position.
DIRECT--a field test of driver information radio systems using experimental communication technology in 30 test vehicles between Detroit and Detroit Metropolitan Airport..
DRIVE I - European system, 1989.
DRIVE II - European system, 1992, (one billion dollar funding).
DRG - Dynamic Route Guidance.
DRMA - Digital Road Map Associaiton (Japan).
DSMA - Data Sense Multiple Access.
ERGS - Electronic Route Guidance Systems; early U.S. program (1968).
ERP - Electronic Road Pricing.
ETTM - Electronic Toll and Traffic Management.

FAR - Federal Acquisition Regulations.

FAST-TRAC--a planned system for faster and safer travel through traffic routing, adaptive traffic signal controls, and dynamic route guidance in Oakland County (Detroit suburbs) Michigan.

FCC - Federal Communications Commission.

FHWA - Federal Highway Administration (U.S.).

GPS - Global Positioning System.

HELP - Heavy vehicle Electronic License Plate; system begun in western North America in 1984 (known as Crescent) for using current technology to facilitate trucking productivity in a crescent-shaped region from British Columbia through California and Arizona to Texas.

HF - Human Factors.

HIDO - Highway Industry Development Organization (Japan).

HOV - High-Occupance Vehicle.

HUD - Head-Up Display.

IHD - Interactive Highway Design Model.

ISTEA - Intermodal Surface Transportation Efficiency Act of 1991.

IVHS Act of 1991 - Intelligent Vehicle-Highway System Act of 1991, a part of the Intermodal Surface Transportation Efficiency Act (ISTEA) of 1991. The thrust of this Act is to better service the customer-traveller, to enhance U. S. business, and to maintain U.S. credibility in this rapidly changing technological arena. Funding priorities are to establish 3 to 10 urban corridors using uniform evaluation crieria. The strategic plan for doing so is then to be implemented using a combination of public and private partnerships with 50 to 80% federal funding.

IVHS -- Advantages

IVHS offers the promise to save lives, avoid injuries, reduce congestion, reduce number of empty trucks, increase travel comfort and convenience, and have a favorable impact on the environment by using less energy than required in current travel systems and in a number of other ways.

IVHS AMERICA (formerly knows as *MOBILITY 2000*)--

This professional organization (1990-) has, among its functions, the role of serving in an advisory capacity to the U.S. Department of Transportation, of developing goals and programs, of fostering research and development, of helping to provide input on decisions concerning IVHS deployment, addressing legal and institutional issues, coordinating cooperation between nations, identifying and developing uniform standards, providing information, alleviating jurisdictional conflict, and determining system architectures.

JSK - Association of Electronics Technology for Automobile Traffic and Driving (Jidosha Soko denshi gijutsu Kyokai).

JTMTA - Japan Traffic Management & Technology Association.

LORAN - Navigation system.

MC - Ministry of Construction (Japan).

MEX - Metropolitan Expressway network (Japan).

MITI - Ministry of International Trade and Industry.

Mobility 2000 (1987) - see IVHS AMERICA.

MOT - Ministry of Transport (Japan).

MPO - Metropolitan Political Organization.

MPT - Ministry of Posts and Telecommunications (Japan).

MT - Ministry of Transport (Japan).

Navigation - system of conceptual principles used to figure position, course and distances, either on the surface of the earth or in the air. Thus, one speaks of navigation with respect to all manner of vehicles: cars, trucks, ships, airplanes, and satellites.

NeGHTS - Next Generation Highway Traffic System, Japanese system; 1989.

NHK - Nippon Hoso Kyokai (Japanese broadcasting system).

NHTSA - National Highway Traffic Safety Administration (U.S.).

NIC - Newly Industrial Country.

NPA - National Police Agency (Japan).

NSF - National Science Foundation.

OBC - On-Board Computers.

Pathfinder - Los Angeles system of traffic information for navigation with 25 test vehicles, funded at $1.7 million for 1990-1992.

Pilotage - refers to the use of visual references to figure position; contrast with navigation.

PROMETHEUS - European system; 1986.

PVS - Personal Vehicle Ssytem (Japan).

RACS - Road Automobile Communication System; Japanese system; 1986.

Radionavigation - navigation based on the radio.

RDS - Radio Data Systems.

RTI - Europe (and Japan along with IVHS).

SCANDI--a field test of surveillance, control, and driver information along 32 freeway miles in Detroit, Michigan; since 1978.

SCATS - Sydney Co-ordinated Adaptive Traffic System; Australian traffic signal control system.

SCOOT - British IVHS.

SDSMA - Switched DSMA.

SMART - Los Angeles demonstration project of 5 arterials covering 14 miles. 1989-1992 funded at $50 million.

SSVS - Super Smart Vehicle System, Japanese system, possibly 1992.

TMT - Japan Traffic Management Technology Association.

TravTek - Orlando, Florida motorist information system covering 1300 square miles; vehicle-based route guidance with 100 test vehicles.

VICS - Vehicle Information and Communications System, Japanese system; 1991.

WIM - Weigh-In-Motion.

RAILROAD TERMINOLOGY GLOSSARY

One vision of the world at the beginning of the third millennium, might include a planet working to build its infrastructure within existing national boundaries and indeed, searching for political solutions to the difficuties in extending elements of the infrastructure across national boundaries. Computerization can free human minds from the mass of detail involved in monitoring and planning such systems, so that they might focus on the deeper abstract philosphical, ethical, and political issues involved in decision making. Within such a world view, digital mapping would be instrumental in freeing human time at managerial levels, and in creating smoothly functioning, accurately planned, spatial layouts for infrastructures. Jobs at the planning level might become even more competitive, with the better positions going to individuals conversant with, or well-trained in, the principles of digital mapping and facilities management. Indeed, at a different scale, even the model train community, which currently uses computer software to plan complicated layouts, might benefit from an understanding of how GIS is becoming integrated into the network of full-sized trains.

The railroad industry in the United States is becoming significant both as a user of GIS and as a component in the transmission of digital information. Fiber optic cable needs to be laid to link long distances and so that direct physical access to it is relatively easy. Using the channels of land devoted to railroad track is one way to gain such access. Further, since 1983, Amtrak (in the United States) has begun using AM/FM and GIS in a real estate capacity to manage the leasing of its land and air rights and in an engineering capacity to manage facilities information such as track layout, various mechanical and electrical systems, and planning. AM/FM and GIS appear to offer substantial promise for innovative management of railway facilities and consequent efficiency of railnet development in the future.

absolute block - a block of defined limits which a train or engine is not permitted to enter while it is occupied by another train or engine.

approach signal - fixed signal used to govern the approach to another signal; if operative, so controlled that its indication furnishes advance information of the indication of the next signal.

automatic block system - a series of consecutive *blocks* governed by block signals, actuated by a train, and engine, or certain conditions affecting the use of a *block*.

block - a length of track of defined limits, the use of which by trains and engines is governed by block signals.

block signal - a fixed signal at the entrance of a block to govern trains and engines entering and using that block.

block system - a series of consecutive blocks.

controlled siding - a siding equipped with control signals which are used to authorize trains or engines to enter or leave the siding; such signals do not govern movements on the siding.

controlled signal - a fixed signal at the entrance of a route or block, controlled manually or otherwise, to govern trains or engines entering and using that route or block.

control station - a place from which a traffic control system or remote control signal appliances are operated or from which instructions are issued.

current of traffic - the movement of trains on a main track, in one direction, specified by rules or time tables.

defect detector - a device at a fixed location which detects dragging equipment or overheated journals and wheels.

division - that portion of a railroad assigned to the supervision of a superintendent.

dwarf signal - a low block or controlled, home, or interlocking signal.

electric lock - an electrical locking device applied to a hand-operated switch or derail.

engine - a unit propelled by any form of energy, or combination of such units operated from a single control, used in train or yard service.

extra train - a train not authorized by time table schedule; it may be designated as: **extra** - an extra train, the movement of which must be in one specified direction; **work extra-** an extra train, the movement of which may be in either direction within specified limits; **passenger extra** - an extra train authorized to operate at passenger train speed.

fixed signal - a signal of fixed location indicating a condition affecting the movement of a train or engine; includes such signals as switch, train order, block, interlocking, stop signs, yard limit signs, slow signs, or other means of displaying indications that govern the movement of a train or engine.

home signal - a fixed signal at the entrance of a route or block used to govern trains or engines entering and using that route or block.

interlocking - an arrangement of signals and signal appliances so interconnected that their movements must succeed each other in proper sequence and for which interlocking rules are in effect; it may be operated manually or automatically.

interlocking limits - the tracks between the opposing home signals of an interlocking.

interlocking signals - the fixed signals of an interlocking.

interlocking station - a place from which an interlocking is operated.

main track - a track, other than an auxiliary track, extending through yards and between stations, upon which trains are operated by time table, train order, or both, or the use of which is governed by block signals or by absolute block authority.

non-signal siding - a siding in signaled territory other than a signaled siding.

non-signaled territory - territory where automatic block systems or traffic control system s are not provided or where systems have been suspended by train order.

permissive block - a block that may be occupied by two or more trains or engines at the same time.

piggyback train - a train heading only piggyback cars (empty piggyback flat cars, trailers, or containers on properly-designed piggyback flat cars or multi-level automobile rack cars); should any other type of freight car be included in the train, it would be designated either an unrestricted or a restricted train, depending upon the equipment or load.

pilot - an employee assigned to a train when the engineman, conductor, or both are not fully acquainted with the physical characteristics or rules of the Railroad, or portion of the Railroad, over which the train is to move.

power-operated switch - a track switch controlled remotely; may be equipped so that it can be operated by hand.

radio - a device used for the transmission of information between moving equipment, between moving equipment and a fixed point, or between fixed points.

radio-equipped - circumstances under which a radio, in position and in condition for use, is provided.

register station - a station at which train register is located.

regular train - a train authorized by a time table schedule.

restricted freight train - a train handling block of thirty or more cars of coal, phosphate, aggregates (including limerock, sand, and so forth) which must be handled next behind the engines, whenever practicable; or one or more restricted cars or loads and which must be operated at designated speed.

run-through switch - a switch which is designed so that it may be trailed through when lined against the desired movement; switches which may be run through in this manner are equipped with switch stands and operating levers that are painted bright yellow; when these switches are trailed through while lined against the movement, a reverse movement must not be made until at least one unit or car has passed beyond the switch points; where snow or sleet is between switch point and stock rail, these switches must not be run through and be lined by hand.

run-through train - a train which passes from one railroad to another with no change in engines or consist other than the addition or removal of a block of one or more cars.

schedule - that part of a time table which prescribes class, direction, number, and movement for a regular train.

section - one of two or more trains running on the same schedule, displaying signals, or for which signals are displayed.

siding - an auxiliary track designated for meeting or passing trains.

signaled siding - a designated siding equipped with block signals governing train and engine movements on the siding.

signaled territory - territory where automatic block systems or traffic control system s are provided and are in operation.

single track - a main track upon which trains are operated in both directions.

speed - **limited speed**, a speed not exceeding 45 miles per hour; **medium speed**, a speed not exceeding 30 miles per hour; **restricted speed**, a speed that will permit stopping within one-half the range of vision, short of train, engine, car, obstruction, broken rail, stop signal, derail or switch not properly lined, but not exceeding 20 miles per hour. **slow speed** - a speed not exceeding 20 miles per hour; **yard speed** - a speed that will permit stopping within one-half the range of vision.

spring switch - a switch equipped with a spring mechanism arranged to restore the switch points to normal positions after having been trailed through.

station - a place designated by name on a time table.

subdivision - a portion of a division designated by a time table.

superior train - a train having precedence over another train.

switch point lock - a locking device installed on certain switch points; where switch point locks are installed part of the throw lever will be painted white.

time table - the authority for the movement of regular trains subject to the rules. It contains classified schedules with special instructions relating to the movement of trains.

Time table Abbreviations: L - Leave; A - Arrive; S - Regular Stop; F - Flag Stop; T - Train Order Office; P - Siding; PN - Northward Siding; PS - Southward Siding; PE - Eastward Siding; PW - Westward Siding; CP - Center Siding; Sig.S - Signaled sidings Of A Capacity In Excess of 250 Cars; X - Railroad Crossing; Y - Wye Track; O - Track Scales; N - North; S - South; E - East; W- West

traffic control system - a block signal system under which train or engine movements are authorized by block signals whose indications supersede the superiority of trains for both opposing and following movements on the same track.

train -an engine, or more than one engine coupled, with or without cars, displaying markers.

train of superior class - a train given precedence by a time table.

train of superior direction - a train given precedence in the direction specified by a time table as between opposing trains of the same class.

train of superior right - a train given precedence by train order.

train register - a book or form which may be used at designated stations for registering signals displayed, the time of arrival and departure of trains, and such other information as may be prescribed.

two or more tracks - two or more train tracks upon any of which the current of traffic may be in either specified direction.

unit run-through train - a run-through train operated by more than one railroad on a continuous round-trip cycle and consisting of assigned equipment.

unrestricted freight train - a train handling regular unrestricted freight cars, both piggyback and box car type equipment or other types of freight cars which are permitted to be operated at maximum freight train speed.

yard - a system of tracks within defined limits provided for the making up of trains, storing cars, and other purposes, over which movements not authorized by time table or by train order may be made (subject to prescribed signals and rules, or special instructions).

yard engine - an engine being used in yard service.

SELECTED REFERENCES

1. **Abler, R. F.** Distance, intercommunications, and geography. *Proceedings of the Association of American Geographers*, 3, 1-4, 1971.
2. **Abler, R. F., Adams, J. S., and Gould, P.** *Spatial Organization*, Prentice-Hall, Englewood Cliffs, New Jersey, 1971.
3. **Ahner, A. L. et al.** Benefit/cost analysis: A business function approach. *Proceedings of the 1990 Annual Conference of the Urban and Regional Information Systems Association II*, 69-80, AUG 1990, URISA Edmonton.
4. **Antenucci, J. C. et al.** *Geographic Information Systems: A Guide to the Technology*, 1991, New York, Van Nostrand Reinhold.
5. **Antenucci, J. and Roitman, H.** Negotiating for system acquisition and maintenance - some legal and practical guidelines. *Proceedings of the 1987 Annual Conference of the Urban and Regional Information Systems Association II*, 25-36, AUG 1987, URISA Fort Lauderdale.
6. **Armstrong, M. P. and Bennett, D. A.** A knowledge based object-oriented approach to cartographic generalization. *GIS/LIS '90 Proceedings 1*, 48-57, 1990, ACSM/ ASPRS /AAG/URISA/AMFM, Anaheim.
7. **Arlinghaus, S. L.** A space for thought in, *Essays on Mathematical Geography*, Institute of Mathematical Geography, Monograph #3, 145-162, Ann Arbor, Michigan, 1986.
8. **Arlinghaus, S. L.** Electronic geometry. *The Geographical Review*, April, 1993, Vol. 83, No. 2, 160-169.
9. **Arlinghaus, S., Arlinghaus, W., and Harary, F.** Sum graphs and geographic information. *Solstice*, 1993, Vol. IV, No. 1. Electronic reference: file SOLSTICE.193, characters 118213-192075, lines 2466-4012. Available via anonymous FTP on account IEVG from site UM.CC.UMICH.EDU.
10. **Arlinghaus, S., Arlinghaus, W., and Harary, F.** *Structural Models in Geography*. Forthcoming.
11. **Aronson, P.** Attribute handling for geographic information systems. *AutoCarto 8 Proceedings Eighth International Symposium on Computer-Assisted Cartography* [Nicholas R. Chrisman, editor], 346-355, APR 1987, ASPRS/ACSM, Baltimore.
12. **Atkin, R. H.** *Mathematical Structure in Human Affairs*, Crane, Russak, New York, 1975.
13. **Austin, R. F.** Raster-vector systems in a production environment. *Proceedings of the 1985 Annual Conference of the Urban and Regional Information Systems Association IV*, 203-211, AUG 1985, URISA, Ottawa.
14. **Austin, R. F.** Praxis as teratism in digital cartography: GIS and AM/FM for the long distance telecommunications industry. *1988 ACSM-ASPRS Fall Convention ACSM Technical Papers*, 198-205, 1988, ACSM/ASPRS, Virginia Beach.
15. **Austin, R. F.** GIS/LIS in the long distance telecommunications industry. *GIS/LIS '88 Proceedings 1*, 266-272, 1988, ACSM/ASPRS/AAG/URISA, San Antonio.
16. **Austin, R. F.** Data bases as the basis for geographic information systems: A perspective. *GIS/LIS '89 Proceedings 1*, 123-133, 1989, ACSM/ ASPRS/ AAG/ URISA/AMFM, Orlando.
17. **Barb, Jr., C. E. and Grace, S. V.** Development of an expert system: The construction bid advisor. *Proceedings of the 1987 Annual Conference of the Urban and Regional Information Systems Association III*, 63-75, AUG 1987, URISA, Fort Lauderdale.

18. **Beard, M. K.** How to survive on a single detailed database. *AutoCarto 8 Proceedings Eighth International Symposium on Computer-Assisted Cartography,* [Nicholas R. Chrisman, editor], 211-220, APR 1987, ASPRS/ACSM, Baltimore.
19. **Bedard, Y.** Uncertainties in land information systems databases. *AutoCarto 8 Proceedings Eighth International Symposium on Computer-Assisted Cartography,* [Nicholas R. Chrisman, editor], 175-184, APR 1987, ASPRS/ACSM, Baltimore.
20. **Bennett, P. G. L.** Identifying and integrating business functions to develop a GIS needs assessment. *Proceedings of the 1990 Annual Conference of the Urban and Regional Information Systems Association III,* 1-13, AUG 1990, URISA, Edmonton.
21. **Berry, J. K.** Maps as data: Fundamental considerations in computer-assisted map analysis. *GIS/LIS '88 Proceedings 1,* 273-284, 1988, ACSM/ASPRS/AAG/URISA, San Antonio
22. **Berry, R. M.** History of geodetic leveling in the United States. *Surveying and Mapping,* 36, 2, 137-153, 1976.
23. **Black Box Corporation.** *Pocket Glossary,* Pittsburgh, PA.
24. **Bossler, J. D.** Necessary data systems. *Proceedings of the International Geographic Information System (IGIS) Symposium '89: Global Systems Science,* 29-30, MAR 1989, AAG/CERMA, Baltimore.
25. **Bragg L. E.** Developing the conventional transverse Mercator projection calculation method. *1986 ASPRS-ACSM Annual Convention Technical Papers 1,* 40-48, MAR 1986, ASPRS/ACSM, Washington.
26. **Brammer, R. F.** A private sector perspective on policy issues related to global system science. *Proceedings of the International Geographic Information System (IGIS) Symposium '89: Global Systems Science,* 23-26, MAR 1989, AAG/CERMA, Baltimore.
27. **Bundock, M. S.** An integrated DBMS approach for geographic information systems. *AutoCarto 8 Proceedings Eighth International Symposium on Computer-Assisted Cartography* [Nicholas R. Chrisman, editor], 292-301, APR 1987, ASPRS/ACSM, Baltimore.
28. **Burrough, P. A.** *Principles of Geographical Information Systems for Land Resources Assessment,* Monographs on Soil and Resources Survey No. 12, Clarendon Press, Oxford, 1986.
29. **Bury, A. S.** Raster to vector conversion: A methodology. *GIS/LIS '89 Proceedings 1,* 9-11, 1989, ACSM/ASPRS/AAG/URISA/AMFM, Orlando.
30. **Cadoux-Hudson, J. and Heywood, D. I.** *Geographic Information 1992/93.* The Yearbook of the Association for Geographic Information. Taylor and Francis, London, 1992.
31. **Carter, J. R.** Digital representations of topographic surfaces: An overview. *1988 ASPRS-ACSM Annual Convention Technical Papers 5,* 54-60, MAR 1988, ASPRS/ACSM, St. Louis.
32. **Carter, J. R.** A typology of geographic information systems. *1988 ASPRS-ACSM Annual Convention Technical Papers 5,* 207-215, MAR 1988, ASPRS/ACSM, St. Louis.
33. **Carter, J. R.** *Computer Mapping Progress in the '80s,* AAG.
34. **Cavo, V. N. and Saylor, C. H. M.** Newly emerging technologies and their impact on AM/FM. *GIS/LIS '89 Proceedings 2,* 794-803, 1989, ACSM/ASPRS/AAG/URISA/AMFM, Orlando.
35. **Chappell, G. B. et al.** A new U.S. Geological Survey data format: Digital line graph-enhanced. *1990 ACSM-ASPRS Fall Convention Technical Papers A,* 9-14, 1990, ACSM/ASPRS, Anaheim.
36. **Chock, M.** The other costs of geographic information systems. *GIS/LIS '90 Proceedings 2,* 526-531, 1990, ACSM/ASPRS/AAG/URISA/AMFM, Anaheim.

37. **Chong, A.** Quality control of stop-and-go kinematic GPS surveying method. *1990 ACSM-ASPRS Fall Convention Technical Papers A,* 114-120, 1990, ACSM/ASPRS, Anaheim.
38. **Chrisman, N. R.** Effective digitizing: advances in software and hardware. *1986 ASPRS-ACSM Annual Convention Technical Papers 1,* 162-171, MAR 1986, ASPRS/ACSM, Washington.
39. **Chrisman, N. R.** Fundamental principles of geographic information systems. *AutoCarto 8 Proceedings Eighth International Symposium on Computer-Assisted Cartography* [Nicholas R. Chrisman, editor], 32-41, APR 1987, ASPRS/ACSM, Baltimore.
40. **Christaller, W.** *Central Places in Southern Germany,* Baskin translation, Prentice-Hall, Englewood Cliffs, NJ, 1966.
41. **Christensen, A. H. J.** Fitting a triangulation to contour lines. *AutoCarto 8 Proceedings Eighth International Symposium on Computer-Assisted Cartography* [Nicholas R. Chrisman, editor], 57-67, APR 1987, ASPRS/ACSM, Baltimore.
42. **Clapp, J. L.** Toward a method for the evaluation of multipurpose land information systems. *Proceedings of the 1986 Annual Conference of the Urban and Regional Information Systems Association I,* 1-10, AUG 1986, URISA.
43. **Clark, S. R. and MacGaffey, N.** Optical disk technology: Potential in geographic information systems, *GIS/LIS '88 Proceedings 2,* 541-549, 1988, ACSM/ASPRS/AAG/URISA, San Antonio.
44. **Clarke, K. C.** *Analytical and Computer Cartography,* Prentice-Hall, Englewood Cliffs, New Jersey, 1990.
45. **Clarke, K. C., et al.** The use of remote sensing and geographic information systems in UNICEF's dracunculiasis (Guinea worm) eradication effort, *Preventive Veterinary Medicine,* 11, 229-235, 1991.
46. **Coates, D. R. and Miller, G. S.** RFP's for cost effective conversion. *Proceedings of the 1990 Annual Conference of the Urban and Regional Information Systems Association II,* 58-68, AUG 1990, URISA, Edmonton.
47. **Codd, E. F.** Relational completeness of data base sublanguages, in *Data Base Systems,* Courant Computer Science Symposia Series, Prentice-Hall, Englewood, New Jersey, 1972.
48. **Codd, E. F.** Relational database: A practical foundation for productivity, *CACM* 25, No. 2, Feb. 1982.
49. **Codd, E. F. and Date, C. J.** Interactive support for nonprogrammers: The relational and network approaches, in C. J. Date, *Relational Database: Selected Writings,* Addison-Wesley, Reading, Massachusetts, 1986.
50. **Community Systems Foundation.** Mapping technologies pack, GeoTECH, 1993, v. 1, no. 1. Prepared for UNICEF by CSF, Ann Arbor, Michigan.
51. **Courington, W.** The network software environment - *A Sun Technical Report,* 0-104, Sun Microsystems, Mountain View, California.
52. **Cowen, D. J.** A versatile cartographic output system for a grid cell GIS. *1987 ASPRS-ACSM Annual Convention Technical Papers V,* 75-81, APR 1987, ASPRS/ACSM, Baltimore.
53. **Cowen, D. J. et al.** Industrial modeling using a geographic information system. *GIS/LIS '90 Proceedings 1,* 1-10, 1990, ACSM/ASPRS/AAG/URISA/AMFM, Anaheim.
54. **Coxeter, H. S. M.** *Non-Euclidean Geometry,* fifth edition, University of Toronto Press, Toronto, 1965.
55. **Coxeter, H. S. M.** *Introduction to Geometry,* John Wiley and Sons, New York, 1961, 1969.
56. **Cross, B. A.** Estimating and comparing tangible and intangible benefits to the costs of geographic information systems. *Proceedings of the 1990 Annual Conference of the*

Urban and Regional Information Systems Association IV, 16-28, AUG 1990, URISA, Edmonton.

57. **Croswell, P. L.** Map accuracy: What is it, who needs it and how much is enough? *Proceedings of the 1987 Annual Conference of the Urban and Regional Information Systems Association II,* 48-62, AUG 1987, URISA, Fort Lauderdale.

58. **Croswell, P. L.** Facing reality in GIS implementation: Lessons learned and obstacles to be overcome. *Proceedings of the 1989 Annual Conference of the Urban and Regional Information Systems Association IV,* 15-35, AUG 1989, URISA.

59. **Croswell, P. L. and Ahner, A.** Computing standards and GIS: A tutorial. *Proceedings of the 1990 Annual Conference of the Urban and Regional Information Systems Association II,* 88-105, AUG 1990, URISA, Edmonton.

60. **Croswell, P. L. and Clark, S. R.** Trends in automated mapping and geographic information system hardware. *1988 ASPRS-ACSM Annual Convention Technical Papers 5,* 69-79, MAR 1988, ASPRS/ACSM, St. Louis.

61. **Dangermond, J.** A technical architecture for GIS. *GIS/LIS '88 Proceedings 2,* 561-570, 1988, ACSM/ASPRS/AAG/URISA, San Antonio.

62. **Dangermond, J.** A review of digital data commonly available and some of the practical problems of entering them into a GIS. *1988 ASPRS-ACSM Annual Convention Technical Papers 5,* 1-10, MAR 1988, ASPRS/ACSM, St. Louis.

63. **Dangermond, J.** Software tools for spatial data integration. *Proceedings of the International Geographic Information System (IGIS) Symposium '89: Global Systems Science,* 31-39, MAR 1989, AAG/CERMA, Baltimore.

64. **Danko, D. M.,** The Digital Chart of the World project. *Proceedings: Technical Papers 1991 ACSM-ASPRS Annual Convention, Volume 2,* 83-93, 1991, ASPRS/ACSM.

65. **Date, C. J.** *An Introduction to Database Systems,* Addison-Wesley, Reading, Massachusett, 1977.

66. **Date, C. J.** A note on the relational calculus. *ACM SIGMOD Record 18,* No. 4, December, 1989.

67. **Date, C. J.** *An Introduction to Database Systems,* fifth edition, Addison-Wesley, Reading, Massachusetts, 1991.

68. **Davis, B. E. and Williams, R.** The five dimensions of GIS. *GIS/LIS '89 Proceedings 1,* 50-58, 1989, ACSM/ASPRS/AAG/URISA/AMFM, Orlando.

69. **Defense Mapping Agency.** *Digital Chart of the World Interim Product Specification,* MIL-D-89009.

70. **Defense Mapping Agency.** *Vector Product Format,* MIL-STD-600006, 1990.

71. **Defense Mapping Agency.** *Interim DCW Tiling Study, 1990.*

72. **Defense Mapping Agency.** *Interim DCW Indexing Study, 1990.*

73. **Dickinson, H. J.** Techniques for establishing the value of geographic information and geographic information systems. *GIS/LIS '89 Proceedings 2,* 412-420, 1989, ACSM/ASPRS/AAG/URISA/AMFM, Orlando.

74. **Digital Geographic Information Working Group.** *Digital Geographic Information Exchange Standards* (DIGEST), final draft, 1989.

75. **Domsch, R. E.** AM/FM/GIS - corridor models, the automation of cross-country facilities. *GIS/LIS '89 Proceedings 1,* 181-186, 1989, ACSM/ASPRS/AAG/URISA/AMFM, Orlando.

76. **Donahue, J. G.** Pure consulting in AM/FM. *1986 ASPRS-ACSM Annual Convention Technical Papers 2,* 97-101, MAR 1986, ASPRS/ACSM, Washington.

77. **Donahue, J. G.** AM/FM - GIS/LIS in the 21st century. *GIS/LIS '89 Proceedings 2,* 520-527, 1989, ACSM/ASPRS/AAG/URISA/AMFM, Orlando.

78. **Duecker, G. T. and Platt, J. T.** The role of automated data checks in the quality assurance of GIS data bases. *GIS/LIS '90 Proceedings 1,* 264-271, 1990, ACSM/ASPRS/AAG/URISA/AMFM, Anaheim.

79. **Dueker, K.** Multi-purpose land information systems: Technical, economic and institutional issues. *AutoCarto 8 Proceedings Eighth International Symposium on Computer-Assisted Cartography* [Nicholas R. Chrisman, editor], 1-11, APR 1987, ASPRS/ACSM, Baltimore.

80. **Dueker, K.** Urban applications of geographic systems technology: A grouping into three levels for resolution. *Proceedings of the 1988 Annual Conference of the Urban and Regional Information Systems Association I*, 104-109, AUG 1988, URISA.

81. **Dugundji, J.** *Topology*, Allyn and Bacon, Boston, Massachusetts, 1966.

82. **Durgin, P. M.** Database guidelines: Are the surveyors and assessors on the same page? *GIS/LIS '89 Proceedings 1*, 342-347, 1989, ACSM/ASPRS/AAG/URISA/ AMFM, Orlando.

83. **Edson, D. T. and Denègre-France, J., eds.** *Glossary of Terms in Computer Assisted Cartography*, International Cartographic Association Commission III, Computer-assisted Cartography, ACSM, Falls Church, Virginia, 1980.

84. **Egenhofer, M. J.** Appropriate conceptual database schema designs for two-dimensional spatial structures. *1987 ASPRS-ACSM Annual Convention Technical Papers V*, 167-179, APR 1987, ASPRS/ACSM Baltimore.

85. **Egenhofer, M. J.** Designing a user interface for a spatial information system. *1988 ASPRS-ACSM Annual Convention Technical Papers 5*, 149-161, MAR 1988, ASPRS/ACSM, St. Louis.

86. **Egenhofer, M. J. and Frank, A. U.** Object-oriented software engineering considerations for future GIS. *Proceedings of the International Geographic Information System (IGIS) Symposium '89: Global Systems Science*, 55-72, MAR 1989, AAG/CERMA, Baltimore.

87. **Eichelberger, P. and Barb, Jr., C. E.** Expert systems and local government: Application areas and future prospects. *Proceedings of the 1987 Annual Conference of the Urban and Regional Information Systems Association III*, 51-62 AUG 1987, URISA, Fort Lauderdale.

88. **Engelhart, P. M. et al.** Implementation of a rapid coordinate transformation device for map projections. *1986 ASPRS-ACSM Annual Convention Technical Papers 1*, 94-103, MAR 1986, ASPRS/ACSM Washington.

89. **Environmental Systems Research Institute, Inc.** *Glossary of GIS and Arc/Info Terms*. Redlands, California, 1991.

90. **Ervin, R. D.** *An American Observation of IVHS in Japan.* Transportation Research Institute, The University of Michigan, Ann Arbor, Michigan, 1991.

91. **Exler, R. D.** Appropriate applications for topologic data structures. *Proceedings of the 1987 Annual Conference of the Urban and Regional Information Systems Association II*, 93-96, AUG 1987, URISA, Fort Lauderdale.

92. **Exler, R.D.** The impact of standards on the GIS industry. *Proceedings of the International Geographic Information System (IGIS) Symposium '89: Global Systems Science*, 191-2050, MAR 1989, AAG/CERMA, Baltimore.

93. **Ferguson, C. W.** Upcoming innovations in conversion technology. *Proceedings of the 1990 Annual Conference of the Urban and Regional Information Systems Association III*, 52-61, AUG 1990, URISA, Edmonton.

94. **Fletcher, D.** Modeling GIS transportation networks. *Proceedings of the 1987 Annual Conference of the Urban and Regional Information Systems Association II*, 84-92, AUG 1987, URISA, Fort Lauderdale.

95. **Foley, J. D. and Van Dam, A.** *Fundamentals of Interactive Computer Graphics.* Addison-Wesley, Reading, Massachusetts, 1982.

96. **Francis, F. J.** Global positioning surveying for a geographic information system. *1986 ASPRS-ACSM Annual Convention Technical Papers 2*, 273-279, MAR 1986, ASPRS/ACSM, Washington.

97. **Frank, A. U.** Overlay processing in spatial information systems. *AutoCarto 8 Proceedings Eighth International Symposium on Computer-Assisted Cartography* [Nicholas R. Chrisman, editor], 13-31, APR 1987, ASPRS/ACSM, Baltimore.

98. **French, R. L.** Automobile navigation in the past, present, and future. *AutoCarto 8 Proceedings Eighth International Symposium on Computer-Assisted Cartography* [Nicholas R. Chrisman, editor], 542-551,APR 1987, ASPRS/ACSM, Baltimore.

99. **Gahegan, M. N. and Roberts, S. A.** An intelligent object-oriented geographical information system. *International Journal of Geographic Information Systems* 2(2), 101-110, 1988.

100. **Garth, S. L.** Developments in GIS data base technologies. *Proceedings of the 1990 Annual Conference of the Urban and Regional Information Systems Association III,* 62-76, AUG 1990, URISA, Edmonton.

101. **Gehring, V.** A multi-participant, public/private county-wide addressing project in Ada County, Idaho. *Proceedings of the 1990 Annual Conference of the Urban and Regional Information Systems Association IV,* 276-287, AUG 1990, URISA, Edmonton.

102. **Gentles, M. E.** What are the secrets to a successful conversion effort? *Proceedings of the 1987 Annual Conference of the Urban and Regional Information Systems Association II,* 37-47, AUG 1987, URISA, Fort Lauderdale.

103. **Gilbert, D. M.** Slivers, gaps and overlaps or precision data entry. *Proceedings of the 1990 Annual Conference of the Urban and Regional Information Systems Association I,* 41-54, AUG 1990, URISA, Edmonton.

104. **Gimblett, R. H. and Itami, R. M.** Linking GIS with video technology to simulate environmental change. *GIS/LIS '88 Proceedings 1,* 208-219, 1988, ACSM/ASPRS/ AAG/URISA, San Antonio.

105. **GIS World, Inc.** *The 1990 GIS Sourcebook,* Fort Collins CO, 1990 and later versions.

106. **Gold, C. M.** Spatial ordering of Voronoi networks and their use in terrain data base management. *AutoCarto 8 Proceedings Eighth International Symposium on Computer-Assisted Cartography* [Nicholas R. Chrisman, editor] 185-194, APR 1987, ASPRS/ACSM, Baltimore.

107. **Goodchild, M. F.** Modeling error in spatial databases. *GIS/LIS '90 Proceedings 1,* 154-162, 1990, ACSM/ASPRS/AAG/URISA/AMFM, Anaheim.

108. **Goodchild, M. and Gopal, S.** *The Accuracy of Spatial Databases,* Taylor and Francis, New York, 1989.

109. **Gorman, J. et al.** Partial shape recognition and orientation estimation. *1988 ASPRS-ACSM Annual Convention Technical Papers 2,* 1-11, MAR 1988, ASPRS/ACSM, St. Louis.

110. **Gould, P.** *Spatial Diffusion,* Association of American Geographers Commission on College Geography, Resource Paper No. 4, Washington, D.C., 1969.

111. **Grady, R. K.** Quality assurance of data: GIGO revisited. *Proceedings of the International Geographic Information System (IGIS) Symposium '89: Global Systems Science,* 181-190, MAR 1989, AAG/CERMA, Baltimore.

112. **Greasley, I.** Data structures to organize spatial subdivisions. *1988 ASPRS-ACSM Annual Convention Technical Papers 5,* 139-148, MAR 1988, ASPRS/ACSM, St. Louis.

113. **Greenlee, D.** Raster and vector processing for scanned linework. *AutoCarto 8 Proceedings Eighth International Symposium on Computer-Assisted Cartography* [Nicholas R. Chrisman, editor], 640-649, APR 1987, ASPRS/ACSM, Baltimore.

114. **Gribb, W. J. and Cherniak, R. J.** Databases and rural addressing geographic information systems. *GIS/LIS '88 Proceedings 2,* 941-947, 1988, ACSM/ASPRS/ AAG/URISA, San Antonio.

115. **Griffith, D. A.** *Spatial Statistics: Past, Present, and Future,* Monograph #12, Institute of Mathematical Geography, Ann Arbor, Michigan, 1990.

116. **Grigg, D.** The logic of regional systems. *Annals of the Association of American Geographers*, 55(3), 1965.
117. **Gugan, D. J. and Hartnall, T. J.** Applications of digital terrain modelling within a GIS. *GIS/LIS '89 Proceedings 2*, 771-780, 1989, ACSM/ASPRS/AAG/URISA/AMFM, Orlando.
118. **Guptill, S. C. et al.** Designing an enhanced digital line graph. *1988 ASPRS-ACSM Annual Convention Technical Papers 2*, 252-261, MAR 1988, ASPRS/ACSM, St. Louis.
119. **Guptill, S. C., ed. et al.** *A Process for Evaluating Geographic Information Systems*, FICCDC Technology Exchange Working Group, Technical Report 1, USGS Open-File Report 88-105, USGS, Reston, Virginia, 1988.
120. **Guthrie, J.** Pages from an address book: Notes on implementing an address database. *Proceedings of the 1987 Annual Conference of the Urban and Regional Information Systems Association I*, 101-110, AUG 1987, URISA, Fort Lauderdale.
121. **Hage, P. and Harary, F.** *Structural Models in Anthropology*, Cambridge University Press, Cambridge, 1983.
122. **Hage, P. and Harary, F.** *Exchange in Oceania: A Graph Theoretic Analysis*, Clarendon Press, Oxford, 1991.
123. **Hage, P. and Harary, F.** *Island Networks*, in press.
124. **Hägerstrand, T.** Aspects of the spatial structure of social communication and the diffusion of information. *Papers and Proceedings of the Regional Science Association*, 16, 27-42, 1965.
125. **Hägerstrand, T.** *Innovation Diffusion as a Spatial Process*, trans. and postscript by Allan Pred, University of Chicago Press, Chicago, 1967.
126. **Hamilton, S. D.** Digital orthophotography and integrated raster/vector geographic information systems: Boondoggle -- or boon? *Proceedings of the 1990 Annual Conference of the Urban and Regional Information Systems Association III*, 124-129, AUG 1990, URISA, Edmonton.
127. **Hamming, R. W.** *Coding and Information Theory*, Prentice-Hall, Englewood Cliffs, New Jersey, 1980.
128. **Harary, F.** *Graph Theory*, 1969, Addison-Wesley, Reading, MA.
129. **Harary, F., Norman, R. Z., and Cartwright, D.** *Structural Models: An Introduction to the Theory of Directed Graphs*, John Wiley and Sons, New York, 1965.
130. **Hazelton, N. W. J. et al.** On the design of temporally-referenced, 3-D geographical information systems: Development of four-dimensional GIS. *GIS/LIS '90 Proceedings 1*, 357-372, 1990, ACSM/ASPRS/AAG/URISA/AMFM, Anaheim.
131. **Henry, B. B.** Problems (and solutions) in building large automated basemaps. *1986 ASPRS-ACSM Annual Convention Technical Papers 2*, 153-158, MAR 1986, ASPRS/ACSM, St. Louis.
132. **Henry, N. F. and Worden, Meredith.** A geographical database design for local emergency service management. *1988 ASPRS-ACSM Annual Convention Technical Papers 5*, 118-124, MAR 1988, ASPRS/ACSM, St. Louis.
133. **Heric, M.** An analysis and comparison of digitized large format camera photography. *1988 ACSM-ASPRS Fall Convention ASPRS Technical Papers 1*, 8, 1988, ACSM/ASPRS, Virginia Beach.
134. **Herring, J. R. et al.** Extensions to the SQL query language to support spatial analysis in a topological data base. *GIS/LIS '88 Proceedings 2*, 741-750, 1988, ACSM/ASPRS/AAG/URISA, San Antonio.
135. **Hill, Q. C.** Tampa (Florida) water meter-reading and route management system. *Proceedings of the 1985 Annual Conference of the Urban and Regional Information Systems Association IV*, 250-257, AUG 1985, URISA, Ottawa.

136. **Hoke, R. E.** Synercom to AutoCAD Translator: The IMAGIS experience. *Proceedings of the 1990 Annual Conference of the Urban and Regional Information Systems Association I*, 241-252, AUG 1990, URISA, Edmonton.

137. **Hoover, M. H.** Managing base map maintenance in a multi-participant GIS project. *Proceedings of the 1990 Annual Conference of the Urban and Regional Information Systems Association IV*, 57-65, AUG 1990, URISA, Edmonton.

138. **Hotelling, H.** Stability in competition. *Economic Journal*, 39, 41-57, 1929.

139. **Hudson, D.** Some comments on data quality in GIS. *1988 ASPRS-ACSM Annual Convention Technical Papers 2*, 203-211, MAR 1988, ASPRS/ACSM, St. Louis.

140. **Hunter, G. J. and Williamson I. P.** The need for a better understanding of the accuracy of spatial databases. *Proceedings of the 1990 Annual Conference of the Urban and Regional Information Systems Association IV*, 120-128, AUG 1990, URISA, Edmonton.

141. **International Cartographic Association.** *WDDES Strategy and Objectives*, 1989.

142. **Jackson, J.** Using Drawing Exchange Format (DXF) files to manage attribute data: A transportation planning example. *Proceedings of the 1990 Annual Conference of the Urban and Regional Information Systems Association III*, 77-86, AUG 1990, URISA, Edmonton.

143. **Ji, M.** TIGER/Line files and their potential for GIS applications. *GIS/LIS '90 Proceedings 1*, 117-124, 1990, ACSM/ASPRS/AAG/URISA/AMFM, Anaheim.

144. **Joffe, B. A.** Guidelines for evaluating AM/GIS systems. *Proceedings of the 1987 Annual Conference of the Urban and Regional Information Systems Association IV*, 222-233, AUG 1987, URISA, Fort Lauderdale.

145. **Kainz, W.** Coordinate transformations in map digitizing. *AutoCarto 8 Proceedings Eighth International Symposium on Computer-Assisted Cartography* [Nicholas R. Chrisman, editor], 107-111, APR, 1987, ASPRS/ACSM, Baltimore.

146. **Karimi, H. et al.** A relational database model for an AVL system and an expert system for optimal route selection. *AutoCarto 8 Proceedings Eighth International Symposium on Computer-Assisted Cartography* [Nicholas R. Chrisman, editor], 584-593, APR 1987, ASPRS/ACSM, Baltimore.

147. **Kelley, J. L.** *General Topology*, D. Van Nostrand, New York, 1955.

148. **Kinnear, C.** The TIGER structure. *AutoCarto 8 Proceedings Eighth International Symposium on Computer-Assisted Cartography* [Nicholas R. Chrisman, editor], 249-257, APR 1987, ASPRS/ACSM Baltimore.

149. **Kolars, J. F. and Malin, H. F.** Population and accessibility: an analysis of Turkish railroads. *Geographical Review*, 60,(2), 229-246, 1970.

150. **Kolars, J. F. and Nystuen, J. D.** Human Geography: Spatial Design in World Society, McGraw-Hill, New York, 1974. Reprinted, Institute of Mathematical Geography, Ann Arbor, MI.

151. **Kjerne, D.** Modeling location for cadastral maps using an object-oriented computer language. *Proceedings of the 1987 Annual Conference of the Urban and Regional Information Systems Association I*, 174-189, AUG 1987, URISA.

152. **Kraklwsky, E. et al.** Research into electronic maps and automatic vehicle location. *AutoCarto 8 Proceedings Eighth International Symposium on Computer-Assisted Cartography* [Nicholas R. Chrisman, editor], 572-583, APR 1987, ASPRS/ACSM, Baltimore.

153. **Kruczynski, L. R.** An introduction to the global position system and its use in urban GIS applications. *Proceedings of the 1990 Annual Conference of the Urban and Regional Information Systems Association III*, 87-91, AUG 1990, URISA, Edmonton.

154. **Kulick, D. J.** Land rights and interests: A brief overview. *1988 ACSM-ASPRS Fall Convention ACSM Technical Papers*, 225-233, 1988, ACSM/ASPRS, Virginia Beach.

155. **Ladha, N. and Robertson, A. G.** Technology and the power line planner. *GIS/LIS '88 Proceedings 2*, 498-506, 1988, ACSM/ASPRS/AAG/URISA, San Antonio.

156. **Lee, J.** A data model for dynamic vehicle routing with GIS. *GIS/LIS '90 Proceedings 1*,134-143, 1990, ACSM/ASPRS/AAG/URISA/AMFM, Anaheim.
157. **Leick, A.** *GPS Satellite Surveying*, 1990, John Wiley and Sons, New York.
158. **Libshitz, A. B.** Data design: A methodology for GIS implementation. *GIS/LIS '89 Proceedings 1*, 243-249, 1989, ACSM/ASPRS/AAG/URISA/AMFM, Orlando.
159. Lindenberg, R. E. and Fisher, P. F. Towards recognition of the functional distinctions among remote sensing, geographic information systems, and cartography. *1988 ACSM-ASPRS Fall Convention ASPRS Technical Papers,* 151-159, 1988, ACSM/ASPRS, Virginia Beach.
160. **Loomer, S. A.** Mathematical analysis of medieval sea charts. *1986 ASPRS-ACSM Annual Convention Technical Papers 1,* 123-132, MAR 1986, ASPRS/ACSM, Washington.
161. **Low, R. M.** Quick-look gridding using quadtree-based interpolation. *1988 ACSM-ASPRS Fall Convention ACSM Technical Papers,* 89-101, 1988, ACSM/ASPRS, Virginia Beach.
162. **Lowe, M. R.** Practical application of structure analysis techniques in a major systems migration effort. *Proceedings of the 1990 Annual Conference of the Urban and Regional Information Systems Association III*, 92-98, AUG 1990, URISA, Edmonton.
163. **MacGaffey, N. and Shalit, H.** Address geocodes in GIS development: Technical considerations from a case study. *GIS/LIS '88 Proceedings 1*, 190-197, 1988, ACSM/ASPRS/AAG/URISA, San Antonio.
164. **Mackaness, W. A. and Fisher, P. F.** Automatic recognition and resolution of spatial conflicts in cartographic symbolisation *AutoCarto 8 Proceedings Eighth International Symposium on Computer-Assisted Cartography* [Nicholas R. Chrisman, editor], 709-718, APR 1987, ASPRS/ACSM, Baltimore.
165. **Mackaness, W. A.** Development of an interface for user interaction in rule based map generalisation. *GIS/LIS '90 Proceedings 1,* 107-116, 1990, ACSM/ASPRS/AAG/URISA/AMFM, Anaheim.
166. **Maddock, B. and Grush, B.** Real-Time interactive registration of digital vector maps. *Proceedings of the 1987 Annual Conference of the Urban and Regional Information Systems Association II,* 63-74, AUG 1987, URISA, Fort Lauderdale.
167. **Madill, R. J.** Mapping and managing utility outside plant records. *Proceedings of the 1985 Annual Conference of the Urban and Regional Information Systems Association IV*, 266-270, AUG 1985, URISA, Ottawa.
168. **Maling, D. H.** *Coordinate Systems and Map Projections,* George Philip and Son Ltd., London, 1973.
169. **Mansfield, M. J.** *Introduction to Topology,* D. Van Nostrand, New York, 1963.
170. **Marble, D. F. and Nystuen, J. D.** An approach to the direct measurement of community mean information fields. *Papers and Proceedings of the Regional Science Association*, 11, 99-109, 1963.
171. **Mark, D. M.** On giving and receiving directions: Cartographic and cognitive issues. *AutoCarto 8 Proceedings Eighth International Symposium on Computer-Assisted Cartography* [Nicholas R. Chrisman, editor], 562-571, APR 1987, ASPRS/ACSM, Baltimore.
172. **Mark, D. M.** Competition for map space as a paradigm for automated map design. *GIS/LIS '90 Proceedings 1,* 97-106, 1990, ACSM/ASPRS/AAG/URISA/AMFM, Anaheim.
173. **Marlow, M. J.** Railroad turnout design and layout. *1988 ASPRS-ACSM Annual Convention Technical Papers 1,* 70-77, MAR 1988 ASPRS/ACSM St. Louis.
174. **Martin, J.** *End-Users' Guide to Data Base,* Prentice-Hall, Englewood Cliffs, New Jersey, 1981.
175. **Martin, J.** *Managing the Data-Base Environment,* Prentice-Hall, Englewood Cliffs, New Jersey, 1983.

176. **McCulloch, T. M. and Marinaro, R. A.** Producing 1:24,000 Scale digital line graphs from scanned data. *1988 ASPRS-ACSM Annual Convention Technical Papers 2,* 243-251, MAR 1988, ASPRS/ACSM, St. Louis.

177. **McGranaghan, M.** Human interface requirements for vehicle navigation aids. *AutoCarto 8 Proceedings Eighth International Symposium on Computer-Assisted Cartography* [Nicholas R. Chrisman, editor], 396-402, APR 1987, ASPRS/ACSM, Baltimore.

178. **McKamey, M. C.** Automated industrial site selection using elevation models and digital image processing, *1987 ASPRS-ACSM Annual Convention Technical Papers V,* 43-52, APR 1987, ASPRS/ACSM, Baltimore.

179. **McMaster, R. B. and Monmonier, M.** A conceptual framework for quantitative and qualitative raster-mode generalization. *GIS/LIS '89 Proceedings 2,* 390-403, 1989, ACSM/ASPRS/AAG/URISA/AMFM, Orlando.

180. **McMaster, R. B. and Shea, K. S.** Cartographic generalization in a digital environment: A framework for implementation in a geographic information system. *GIS/LIS '88 Proceedings 1,* 240-249, 1988, ACSM/ASPRS/AAG/URISA, San Antonio.

181. **McRae, S. D.** GIS design and the questions users should be asking, *GIS/LIS '89 Proceedings 2,* 528-537, 1989, ACSM/ASPRS/AAG/URISA/AMFM, Orlando.

182. **Miller, M. W.** U.S. Spies Help Scientists Pierce Data Jungle. *The Wall Street Journal,* B1, July 27, 1993.

183. **Moellering, H. (Ed.)** *Spatial Database Transfer Standards: Current International Status,* ICA, Elsevier, New York, 1991.

184. **Monmonier, M. S.** *Computer-Assisted Cartography: Principles and Prospects,* Prentice-Hall, Englewood Cliffs, New Jersey, 1982.

185. **Monmonier, M. S.** *How to Lie with Maps,* University of Chicago Press, Chicago, 1991.

186. **Monmonier, M. S.** *Mapping it Out: Expository Cartography for the Humanities and Social Sciences,* University of Chicago Press, Chicago, 1993.

187. **Montgomery, G. E.** Multiparticipant projects - Achieving consensus on technical and funding cost allocation issues. *Proceedings of the 1987 Annual Conference of the Urban and Regional Information Systems Association IV,* 204-211, AUG 1987, URISA, Fort Lauderdale.

188. **Moreland, D. K.** Development of an automated laser-based data capture system at the U.S. Geological Survey. *1986 ASPRS-ACSM Annual Convention Technical Papers 1,* 189-198, MAR 1986, ASPRS/ACSM, Washington.

189. **Mower, J. E.** A neural network approach to feature recognition along cartographic lines. *GIS/LIS '88 Proceedings 1,* 250-255, 1988, ACSM/ASPRS/AAG/URISA, San Antonio.

190. **Moyer, D. D.** Comparing the costs: Manual versus automated procedures for handling land records. *1988 ASPRS-ACSM Annual Convention Technical Papers 5,* 198-206, MAR 1988, ASPRS/ACSM, St. Louis.

191. **National Research Council,** *Procedures and Standards for a Multipurpose Cadastre,* National Academy Press, Washington DC, 1983.

192. **Ness, G. D.; Drake, W. D.; and Brechin, S. R.; eds.** *Population-Environment Dynamics: Ideas and Observations,* The University of Michigan Press, Ann Arbor, Michigan, 1993.

193. **Nordisk K.** Joint Nordic Project Community benefit of digital spatial information. *Report 3 Digital Map Data Bases Economics and User Experiences in North America,* ISBN 951-47-0366-9, 1987.

194. **Nystuen, J. D.** Identification of some fundamental spatial concepts. *Papers of the Michigan Academy of Science, Arts and Letters,* 48, 373-384, 1963.

195. **Nystuen, J. D. Ed.** *Michigan Inter-university Community of Mathematical Geographers, Papers, 1963-1968,* Ann Arbor, MI. Reprinted, Institute of Mathematical Geography, Ann Arbor, Michigan.

196. **O'Connor, K. and Sidebottom, G.** A cartographic document interpretation system. *Proceedings of the 1990 Annual Conference of the Urban and Regional Information Systems Association II,* 256-268, AUG 1990, URISA, Edmonton.

197. **Oaten, G. W. and Shortis, M. R.** An object-oriented approach to analysis for geographic information systems. *Proceedings of the 1990 Annual Conference of the Urban and Regional Information Systems Association II,* 244-255, AUG 1990, URISA, Edmonton.

198. **Obermeyer, N. J.** A systematic approach to the taxonomy of geographic information use. *GIS/LIS '89 Proceedings 2,* 421-429, 1989, ACSM/ASPRS/AAG/URISA/AMFM, Orlando.

199. **Onsrud, H. J.** Understanding the uses and assessing the value of geographic information. *GIS/LIS '89 Proceedings 2,* 404-411, 1989, ACSM/ASPRS/AAG/URISA/AMFM, Orlando.

200. **Openshaw, S.** Learning to live with errors in spatial databases, in *The Accuracy of Spatial Data Bases,* edited by Michael Goodchild and Sucharita Gopal, 263-276, Taylor and Francis, New York, 1989.

201. **Oshima, T.** Computer aided route selection on the base of GIS by a micro-computer. *1986 ASPRS-ACSM Annual Convention Technical Papers 3,* 31-37, MAR 1986, ASPRS/ACSM, Washington.

202. **Padmanabhan, G., Yoon, J., and Leipnik, M.** *A Glossary of GIS Terminology,* Technical Report 92-13, NCGIA, Santa Barbara, California, December, 1992.

203. **Palmer, D.** Designing a GIS/LIS: Some accuracy and cost considerations. *Proceedings of the 1989 Annual Conference of the Urban and Regional Information Systems Association II,* 52-56, AUG 1989, URISA.

204. **Parent, P.** Application development in the evolution of geographic information systems. *1988 ACSM-ASPRS Fall Convention ASPRS Technical Papers,* 165-174, 1988, ACSM/ASPRS, Virginia Beach.

205. **Parent, P.** How to write a RFP for consultant services for geographic information system implementation. *Proceedings of the 1990 Annual Conference of the Urban and Regional Information Systems Association II,* 52-57, AUG 1990, URISA, Edmonton.

206. **Parent, P. et al.** Estimating the costs of building your AM/GIS database. *GIS/LIS '89 Proceedings 1,* 143-151, 1989, ACSM/ASPRS/AAG/URISA/AMFM, Orlando.

207. **Petersen, J. K.** An updating procedure for digital map databases. *GIS/LIS '89 Proceedings 1,* 134-142, 1989, ACSM/ASPRS/AAG/URISA/AMFM, Orlando.

208. **Pfaffenberger, B.** *Que's Computer User's Dictionary,* Que Corporation, Carmel, Indiana, 1990.

209. **Philbrick, A. K.** *This Human World,* John Wiley and Sons, New York, 1963. Reprinted by Institute of Mathematical Geography, Ann Arbor, Michigan.

210. **Plumb, G. A.** Displaying GIS data sets using cartographic classification techniques. *GIS/LIS '88 Proceedings 1,* 340-349, 1988, ACSM/ASPRS/AAG/URISA, San Antonio.

211. **Poor, A. Where Am I?** *PC Magazine,* 31, November 24, 1992.

212. **Pressman, R. S.** *Software Engineering,* 2nd edition, McGraw-Hill, New York, 1987.

213. **Pullar, D. V.** Query language for spatial relationships. *1987 ASPRS-ACSM Annual Convention Technical Papers V,* 180-192, APR, 1987, ASPRS/ACSM, Baltimore.

214. **Pullar, D. V.** Data definition and operators on a spatial data model, *1988 ASPRS-ACSM Annual Convention Technical Papers 2,* 196-202, MAR 1988, ASPRS/ACSM, St. Louis.

215. **Rector, J. M.** An AM/FM success story: Southern Belltion Systems, *GIS/LIS '90 Proceedings 2,* 541-546, 1990, ACSM/ASPRS/AAG/URISA/AMFM, Anaheim.

216. **Rinehart, Robert E. and Coleman, Earl J.** Digital elevation models produced from digital line graphs. *1988 ASPRS-ACSM Annual Convention Technical Papers 2*, 291-299, MAR 1988, ASPRS/ACSM, St. Louis.

217. **Ripple, W. J., ed.** *GIS for Resource Management, A Compendium,* American Society for Photogrammetry and Remote Sensing, Falls Church, Virginia, 1987.

218. **Robinson, A. H., Sale, R. D., Morrison, J. L., and Muehrcke, P. C.** *Elements of Cartography*, fifth edition, John Wiley and Sons, New York, 1984.

219. **Robinson V. B. and Frank, A.** Expert systems applied to problems in geographic information systems: Introduction, review and prospects. *AutoCarto 8 Proceedings Eighth International Symposium on Computer-Assisted Cartography* [Nicholas R. Chrisman, editor], 396-402, APR 1987, ASPRS/ACSM, Baltimore.

220. **Roitman, H.** Legal issues in providing access to an AMS: Case studies and variances. *Proceedings of the 1987 Annual Conference of the Urban and Regional Information Systems Association II*, 13-24, AUG 1987, URISA, Fort Lauderdale.

221. **Rosen, K. H.** *Discrete Mathematics and its Applications*, Random House, New York, 1988.

222. **Rosen, K. H. and Michaels, J. G., eds.** *Applications of Discrete Mathematics*, McGraw-Hill, New York, 1991.

223. **Rosen, K. H. et al.** *UNIX System V Release 4: An Introduction for New and Experienced Users*, Osborne McGraw-Hill, Berkeley, 1990.

224. **Rossmeissl, H. J.** The spatial data transfer standard: A progress report. *GIS/LIS '89 Proceedings 2*, 699-706, 1989, ACSM/ASPRS/AAG/URISA/AMFM, Orlando.

225. **Sakashita, S. and Tanaka, Y.** Computer-aided drawing conversion (an interactive approach to digitize maps). *GIS/LIS '89 Proceedings 2*, 578-590, 1989, ACSM/ASPRS/AAG/URISA/AMFM, Orlando.

226. **Seaborn, D. W.** Distributed processing and distributed databases in GIS - Separating hype from reality. *GIS/LIS '88 Proceedings 1*, 141-144, 1988, ACSM/ASPRS/ AAG/ URISA, San Antonio.

227. **Selden, D. D.** Automated cartographic data editing: A method for testing and evaluation. *1986 ASPRS-ACSM Annual Convention Technical Papers 1*, 5-14, MAR 1986, ASPRS/ACSM, Washington.

228. **Shea, K. S.** Cartographic data entry through automatic feature tracking. *AutoCarto 8 Proceedings Eighth International Symposium on Computer-Assisted Cartography* [Nicholas R. Chrisman, editor], 660-669, APR 1987, ASPRS/ACSM, Baltimore.

229. **Simmons, Al. Editor.** *Government Technology Resource Guide for GIS,* Sacramento, California.

230. **Sipp, E. N.** Solving public utility problems by integrating GIS with distribution analysis systems. *Proceedings of the 1990 Annual Conference of the Urban and Regional Information Systems Association I*, 231-240, AUG 1990, URISA, Edmonton.

231. **Skiles, James M.** Tools and techniques used in scan conversion of geographic and facilities data. *GIS/LIS '88 Proceedings 1*, 58-67, 1988, ACSM/ASPRS/AAG/URISA, San Antonio.

232. **Skiles, J. M.** Using scanning, automated conversion, and raster data to speed GIS implementation and lower cost. *GIS/LIS '90 Proceedings 2*, 476-483, 1990, ACSM/ASPRS/AAG/URISA/AMFM, Anaheim.

233. **Sloane, N. J. A. and MacWilliams, F. J.** *The Theory of Error Correcting Codes*, Elsevier/North Holland, New York, 1983.

234. **Smith, D. R.** Surface modeling in site selection. *1988 ASPRS-ACSM Annual Convention Technical Papers 5*, 48-53, MAR 1988, ASPRS/ACSM, St. Louis.

235. **Somers, R. and Eichelberger, P.** Development of an integrated cadastral database. *Proceedings of the 1987 Annual Conference of the Urban and Regional Information Systems Association II*, 75-83, AUG 1987, URISA, Fort Lauderdale.

236. **Speer, T.** CAD to GIS: A study of application evolution. *1988 ACSM-ASPRS Fall Convention ACSM Technical Papers,* 217-224, 1988, ACSM/ASPRS, Virginia Beach.

237. **Spencer, R. and Menard, R. D.** Integrating raster/vector technology in a geographic information system. *GIS/LIS '89 Proceedings 1,* 1-8, 1989, ACSM/ASPRS/AAG/URISA/AMFM, Orlando.

238. **Star, J. and Estes, J.** *Geographic Information Systems: An Introduction,* Prentice-Hall, Englewood Cliffs, New Jersey, 1990.

239. **Stockton, B.** Integrating GIS within a corporate database management strategy. *Proceedings of the 1990 Annual Conference of the Urban and Regional Information Systems Association III,* 28-37, AUG 1990, URISA, Edmonton.

240. **Stow, D. et al.** Raster-vector integration for updating land use data. *Proceedings of the 1990 Annual Conference of the Urban and Regional Information Systems Association IV,* 256-264, AUG 1990, URISA, Edmonton.

241. **Thorpe, J. A.** Contour interpolation in large scale urban mapping systems. *1988 ASPRS-ACSM Annual Convention Technical Papers 5,* 61-68, MAR 1988, ASPRS/ACSM, St. Louis.

242. **Thünen, J. H. von.** *The Isolated State,* Wartenberg trans; edited with an introduction by Peter Hall. Oxford, Pergamon Press, New York, 1966 [1842].

243. **Tobler, W. R.** *Map Transformations of Geographic Space,* Doctoral Dissertation, University of Washington, 1961.

244. **Tobler, W. R.** Automation and cartography. *The Geographical Review,* 49, 526-534, 1959.

245. **Tobler, W. R.** An experiment in the computer generalization of maps. *Department of Geography Technical Report #1,* University of Michigan, Ann Arbor, 1964.

246. **Tobler, W. R.** Geographical coordinate computations. *Department of Geography Technical Reports #2 and 3,* University of Michigan, Ann Arbor, 1964.

247. **Tobler, W. R.** Satellite confirmation of settlement size coefficients. *Area I,* 3, 31-34, 1969.

248. **Tobler, W. R.** *Selected Computer Programs,* Michigan Geographical Publications, #1, Department of Geography, University of Michigan, Ann Arbor, 1970.

249. **Tobler, W. R.** The hyperelliptical and other new pseudo cylindrical equal area map projections. *Journal of Geophysical Research,* 78(11), 1753-1759, 1973.

250. **Tobler, W. R.** Analytical cartography. *The American Cartographer,* 3(1,) 21-31, 1976.

251. **Tobler, W. R.** Smooth pycnophylactic interpolation for geographical regions. *Journal of the American Statistical Association,* 74, 519-530, 1979.

252. **Tobler, W. R.** GIS Transformations. *GIS/LIS '90 Proceedings1,* 163-166, 1990, ACSM/ASPRS/AAG/URISA/AMFM, Anaheim.

253. **Tschudi, M. K.** New life for map videodics. *GIS/LIS '90 Proceedings 2,* 547-554, 1990, ACSM/ASPRS/AAG/URISA/AMFM, Anaheim.

254. **Unwin, D. J. and Dawson, J. A.** *Computer Programming for Geographers,* Longman, London, 1985.

255. **Van Biljon, W.** A geographical database system. *AutoCarto 8 Proceedings Eighth International Symposium on Computer-Assisted Cartography* [Nicholas R. Chrisman, editor], 356-362, APR 1987, ASPRS/ACSM, Baltimore.

256. **Van Roessel, J. W.** Design of a spatial data structure using relational normal forms, *International Journal of Geographic Information Systems,* 1(1), 33-50, 1987.

257. **Van Roessel, J. W.** An algorithm for locating candidate labeling boxes with a polygon. *AutoCarto 8 Proceedings Eighth International Symposium on Computer-Assisted Cartography* [Nicholas R. Chrisman, editor], 689-700, APR 1987, ASPRS/ACSM, Baltimore.

258. **Vanzella, L. and Cabay, S.** Hybrid spatial data structures. *GIS/LIS '88 Proceedings 1,* 360-372, 1988, ACSM/ASPRS/AAG/URISA, San Antonio.

259. **Ventura, S. J.** Framework for evaluating GIS implementation. *GIS/LIS '89 Proceedings 2*, 825-836, 1989, ACSM/ASPRS/AAG/URISA/AMFM, Orlando.

260. **Vraga, R. S.** Paneling and partitioning for a tiled data base. *1988 ACSM-ASPRS Fall Convention ACSM Technical Papers,* 157-161, 1988, ACSM/ASPRS, Virginia Beach.

261. **Wagner, D. F.** A method of evaluating polygon overlay algorithms. *1988 ASPRS-ACSM Annual Convention Technical Papers 5*, 173-183, MAR 1988, ASPRS/ACSM, St. Louis.

262. **Wallace, T. and Clark, S. R.** Raster and vector data integration: Past techniques, current capabilities, and future trends. *GIS/LIS '88 Proceedings 1*, 418-426, 1988, ACSM/ASPRS/AAG/URISA, San Antonio.

263. **Ward, B.** User-oriented format for topological vector data bases. *GIS/LIS '90 Proceedings 1,* 402-412, 1990, ACSM/ASPRS/AAG/URISA/AMFM, Anaheim.

264. **Watson, Jr., C. C.** Portable GIS: Using and maintaining a GIS data base in the real world *Proceedings of the 1990 Annual Conference of the Urban and Regional Information Systems Association II*, 145-151, AUG 1990, URISA, Edmonton.

265. **Waugh, T. C. and Buttenfield, B. P.** The GEOVIEW Design: A relational data base approach to geographic data handling. *International Journal of Geographic Information Systems*, 1,(2), 101-118; *GIS/LIS '88 Proceedings 1*, 350-359, 1988, ACSM/ASPRS/AAG/URISA, San Antonio.

266. **Westerfeld, F. E.** Transportation system analysis for emergency services operations. *Proceedings of the 1987 Annual Conference of the Urban and Regional Information Systems Association IV*, 132-141, AUG 1987, URISA, Fort Lauderdale.

267. **Westmoreland, S. J. and Stow, D.** Use of satellite imagery and ancillary data to update a vector-coded geographic information system. *GIS/LIS '90 Proceedings 1,* 383-391, 1990, ACSM/ASPRS/AAG/URISA/AMFM, Anaheim.

268. **White, Jr., M. S.** Technical requirements and standards for a multipurpose geographic data system. *The American Cartographer*, 11(1), 15-26, 1984.

269. **White, Jr., M. S.** Digital map requirements of vehicle navigation. *AutoCarto 8 Proceedings Eighth International Symposium on Computer-Assisted Cartography* [N. R. Chrisman, editor], 552-561, APR 1987, ASPRS/ACSM, Baltimore.

270. **Williams, J. W. and Terp, J. A.** *DigiTech Systems' Dictionary*, Spring 1991, Indianapolis, Indiana, Spring, 1991.

271. **Wu, S-T.** A digital video and image analysis system for GIS/LIS. *GIS/LIS '89 Proceedings 2*, 727-732, 1989, ACSM/ASPRS/AAG/URISA/AMFM, Orlando.

272. **Wunneburger, Douglas F. et al.** Generating automated digital mosaics from aerial videography using georeferencing and stereo image mapping. *1990 ACSM-ASPRS Fall Convention Technical Papers B*, 196-203, 1990, ACSM/ASPRS, Anaheim.

273. **Xiao, Q. et al.** A temporal/spatial database structure for remotely sensed image data management within GIS. *GIS/LIS '89 Proceedings 1*, 116-122, 1989, ACSM/ASPRS/AAG/URISA/AMFM, Orlando.

274. **Young, O. L.** Planning for quality control and managing for quality assurance. *GIS/LIS '89 Proceedings 2*, 629-635, 1989, ACSM/ASPRS/AAG/URISA/AMFM, Orlando.

275. **Zeller, M. N.** Complete translation of graphic and attribute cartographic information between computer file formats. *1987 ASPRS-ACSM Annual Convention Technical Papers V*, 141-150, APR 1987, ASPRS/ACSM, Baltimore.

276. **Zeller, M. N.** On the selection of GIS software for electric utilities. *GIS/LIS '88 Proceedings 2*, 628-638, 1988, ACSM/ASPRS/AAG/URISA, San Antonio.

277. **Zetlan, A. J.** The Methodology of Prototyping: Shortcut to Success. *GIS/LIS '89 Proceedings 2*, 620-628, 1989, ACSM/ASPRS/AAG/URISA/AMFM, Orlando.

278. **Zoraster, S.** Manual and automated line generalization and feature displacement. *U.S. Army Engineer Topographic Laboratories Report ETL-0359*, US Army E.T.L. Virginia, 1984.

279. **Zoraster, S. and Bayer, S.** Practical experience with a map label placement algorithm. *AutoCarto 8 Proceedings Eighth International Symposium on Computer-Assisted Cartography* [Nicholas R. Chrisman, editor], 701-708, APR 1987, ASPRS/ACSM, Baltimore.

PART III: INDICES
INDEX TO THE CASE STUDY
Cross-referenced with the Terms and Concepts Part

*NUMBERS **BELOW** REFER TO **PAGE** NUMBERS IN THE BOOK*

The following set of words from the case study also appear in Part I of this Handbook. Thus, an individual who looks up a word in Part I, might then look in this index to find out where the term or concept was used in an actual case study.

INDEX TO THE PLATES
Cross-referenced with the Terms and Concepts Part
and with the Case Study.

Plates XII through XVI refer to the Case Study and are directly referred to in it. The list below refers to Plates I through XI and is cross-referenced with the alphabetized Terms and Concepts of Part I. First, each Plate is listed, and the set of terms in which it is mentioned in Part I is listed; this strategy gives the reader a feeling for what sorts of terms and concepts might be associated with each plate. Then, the entire list (for all eleven plates) is sorted alphabetically as a single list to give the reader a feel for the comprehensive scope of ideas even a small set of images can suggest.

Thus, an individual who looks up a word in Part I, might then look in this index to find out where the term or concept appears in a graphic context. This index is designed to offer deeper practical understanding of terms and concepts by pointing out where they have been used in the world of digital images. Similarly, an individual attracted to a particular plate can look up terms, in the alphabetically ordered part on terms and concepts, to learn more about it.

model
orthographic perspective
pin diagram
pit
pour point
relief
scene generation
shaded relief map
solid modeling
surface mapping
topographic features
viewshed
watershed
Plate X.
buffer zone

configuration
contiguity analysis
geographic calibration
isoline
nearest neighbor analysis
polygons
polygon overlay
Plate XI.
color
cultural features
discrete
line artwork
line mapping
network
rate map

Entire list sorted alphabetically

accessibility
adjoining sheets
advanced visible/infrared imaging
           spectrometer imaging
aerial photography
agglomeration
albedo
alignment
allocation
analytical hill shading
analytical triangulation
attitude
attribute
automated mapping/facilities management
backdrop
base map
basic cover
block diagram
boundary
buffer zone
building cable
cable marker
cadastral map
calibration
camera
choropleth
clarity
clip
clump
coastline
color

color infrared photography
common profile
compilation (map)
composite
conditional map element
configuration
contiguity analysis
contiguous
continuing property record
contour
contrast
coordinates
coverage
cross section
cultural features
data spacing
degree
depression contour
derived map
digital contour plot
digital elevation model
digital terrain model
dimension
discrete
dissolve lines
distance
drain line
drape
easement
edge
element

INDEX TO THE REFERENCES

Cross-referenced with the Terms and Concepts Part

*NUMBERS **BELOW** REFER TO NUMBERED **CITATIONS** IN THE REFERENCES*

The following set of words from the references also appear in Part I of this Handbook. Thus, an individual who looks up a word in Part I, might then look in this index to find out where the term or concept was used in a research or educational context. This index is designed to offer deeper practical understanding of terms and concepts by pointing out where they have been used in the research world.

The search was done in context on a computer. Thus, not all instances of the word "format" that appear in the reference set have their citation number written next to the index entry for "format"--otherwise, the search would include "trans*format*ion" and "in*format*ion" under "format." Blank spaces can be inserted to overcome this difficulty, but then, "formatting" is ignored in the search for "format." Thus, a search in context was conducted.